Assessment Tests for Higher Chemistry

D A Buchanan
(Moray House School of Education, University of Edinburgh)

J R Melrose
(formerly Lenzie Academy)

Published by
Chemcord
Inch Keith
East Kilbride
Glasgow

ISBN 9781870570985

© Buchanan and Melrose, 2015

All rights reserved. No part of this publication may be reproduced or transmitted in any form or by any means, electronic or mechanical, including photocopy, recording, or any information storage and retrieval system, without permission in writing from the publisher or under licence from the Copyright Licensing Agency.

Printed by Bell and Bain Ltd, Glasgow

Note to teachers / students

The tests are specifically designed to pin-point areas of difficulty and to check students' understanding of all of the essential content in the Higher course.

While the tests can be administered to the whole class, it is suggested that they can be more effectively used by students working at their own pace in class, during self-study time in school or as homework. The information from the results of the tests can be used to help students to plan revision. The test results can also be used by teachers / lecturers who are interested in assessing individual or class difficulties.

Each test is, by and large, independent of the others and consequently the tests can be used to fit almost any teaching order.

The variation in length of the tests is a reflection of the different kinds of question that are associated with a particular area of content. Consequently, different allocations of time are required.

Some questions refer to the Data Booklet. This can be downloaded from the SQA website.
http://www.sqa.org.uk/sqa/files_ccc/H_SQP_Chemistry-Data-Booklet-DQP.pdf

Questions in "The Avogadro Constant (ii)" on page 100 are not part of the mandatory content of the course. However, given the Avogadro Constant, students could perhaps meet calculations of this kind in a problem-solving context.

A complete set of answers are included at the back of the book.

Acknowledgement

A number of questions in the tests come from or have evolved from questions used in Scottish Qualifications Authority (SQA) examinations. The publisher wishes to thank the SQA for permission to use the examination questions in this way.

Index

Chemical Changes and Structure	Test 1.1	Collision theory and activation energy	1
	Test 1.2	Enthalpy changes	4
	Test 1.3	Reaction profiles	6
	Test 1.4	Bonding and structure (i)	13
	Test 1.5	Forms of carbon	14
	Test 1.6	Trends in the Periodic Table (i)	15
	Test 1.7	Trends in the Periodic Table (ii)	17
	Test 1.8	Ionisation energy	19
	Test 1.9	Bonding and structure (ii)	22
	Test 1.10	Bonding and structure (iii)	24
	Test 1.11	Bonding and structure (iv)	25
	Test 1.12	Properties of elements and compounds (i)	27
	Test 1.13	Properties of elements and compounds (ii)	29
	Test 1.14	Polarity of molecules	32
	Test 1.15	Problem solving	33
Nature's Chemistry	Test 2.1	Structure of hydrocarbons (revision)	42
	Test 2.2	Systematic naming of hydrocarbons (revision)	46
	Test 2.3	Isomeric hydrocarbons (revision)	48
	Test 2.4	Functional groups (i)	50
	Test 2.5	Simple esters – structures and names	51
	Test 2.6	Isomers again	52
	Test 2.7	Addition reactions (revision)	54
	Test 2.8	Reactions involving alcohols	58
	Test 2.9	Structure of alcohols	60
	Test 2.10	Oxidation (i)	61
	Test 2.11	Oxidation (ii)	63
	Test 2.12	Simple esters – properties and reactions	66
	Test 2.13	Organic reactions	67
	Test 2.14	Addition polymers (revision)	69
	Test 2.15	Condensation polymers	72
	Test 2.16	Fats and oils	76
	Test 2.17	Soaps, detergents and emulsifiers	79
	Test 2.18	Proteins	82
	Test 2.19	Functional groups (ii)	86
	Test 2.20	Everyday chemistry	87
	Test 2.21	Terpenes	90
	Test 2.22	Free radicals	92

Chemistry in Society

Test 3.1	The mole (revision)	94
Test 3.2	More on the mole	95
Test 3.3	The Avogadro Constant (i)	98
Test 3.4	The Avogadro Constant (ii)	100
Test 3.5	Molar volume	102
Test 3.6	Calculations based on equations (i)	104
Test 3.7	The idea of excess	107
Test 3.8	Calculations based on equations (ii)	109
Test 3.9	Equilibrium (i)	113
Test 3.10	Equilibrium (ii)	116
Test 3.11	Equilibrium (iii)	118
Test 3.12	Percentage yield	122
Test 3.13	Atom economy	123
Test 3.14	Enthalpy of combustion	124
Test 3.15	Hess's Law	126
Test 3.16	Bond enthalpy	130
Test 3.17	Oxidation and reduction (revision)	133
Test 3.18	Oxidising and reducing agents	134
Test 3.19	Redox reactions	135
Test 3.20	Writing ion-electron equations	136
Test 3.21	Neutralisation (revision)	137
Test 3.22	Redox titrations	139
Test 3.23	Chromatography	141
Test 3.24	Practical skills	145

Answers 148

Test 1.1   Collision theory and activation energy

In questions 1 to 12 decide whether each of the statements is

A.   TRUE   B.   FALSE.

1. The collision theory can be used to explain the effect of concentration on reaction rates.

2. The collision theory **cannot** be used to explain the effect of particle size on reaction rates.

3. The activation energy is the minimum kinetic energy required by colliding particles before reaction will occur.

4. The increase in reaction rate with increasing temperature **cannot** just be explained by an increase in the number of collisions.

5. The effect of temperature on reaction rates can be explained in terms of an increase in the size of particles.

6. Energy distribution diagrams can be used to explain the effect of changing temperature on reaction rates.

7. The effect of temperature on reaction rates can be explained in terms of an increase in the number of particles with energy greater than the activation energy.

8. For some reactions to occur, the particles must have the correct collision geometry.

9. All particles at the same temperature have the same kinetic energy.

10. A catalyst lowers the energy that molecules need for successful collisions.

11. A catalyst provides energy so that more molecules have successful collisions.

12. A catalyst decreases the activation energy for a reverse reaction.

13. The activation energy for a forward reaction is the energy difference between the reactants and the products.

14. For any chemical, the temperature is a measure of
    A. the average kinetic energy of the particles that can react
    B. the average kinetic energy of all the particles
    C. the activation energy required for a reaction to occur
    D. the minimum kinetic energy required for a reaction to occur.

15. The same reaction was carried out at four different temperatures. The table shows the times taken for the reaction to occur.

Temperature / °C	20	30	40	50
Time / s	60	30	14	5

    The results show that
    A. the activation energy increases with increasing temperature
    B. the rate of the reaction is directly proportional to the temperature
    C. a small rise in temperature results in a large increase in reaction rate.

16.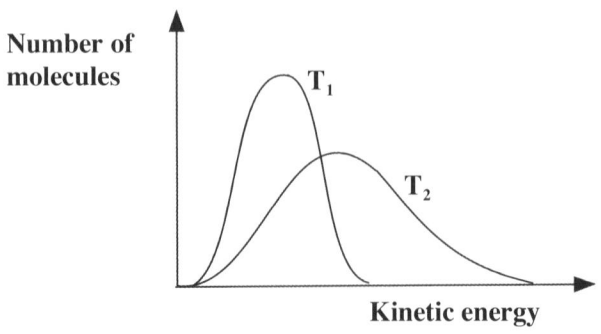

    Which line in the table shows the effect of increasing the temperature of a reaction from T_1 to T_2?

	Activation energy (Ea)	Number of successful collisions
A.	remains the same	increases
B.	decreases	decreases
C.	decreases	increases
D.	remains the same	decreases

Chemical Changes and Structure

17. The graph opposite shows the distribution of kinetic energies of the molecules in a sample of gas.

    Which graph would show the kinetic energies of the molecules when the sample is cooled by 10 °C?

    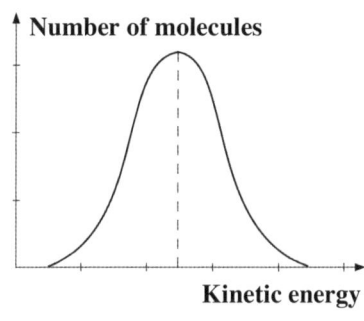

    A.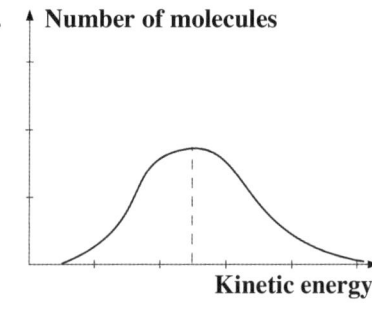

    B.

    C.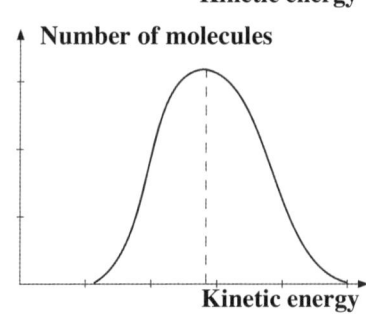

    D.

18. A small increase in temperature results in a large increase in rate of reaction.

    The **main** reason for this is that

    A. more collisions are taking place
    B. the activation energy is lowered
    C. many more particles have energy greater than the activation energy.

Chemical Changes and Structure

Test 1.2　　　　　　　　　　　　Enthalpy changes

1. What types of changes cause heat to be released to the surroundings?

    A. exothermic　　　　　　**B.** endothermic

2. What types of changes cause heat to be taken in from the surroundings?

    A. exothermic　　　　　　**B.** endothermic

3. If the total energy change for the bond breaking steps is less than for the bond making steps, the overall reaction will be

    A. exothermic　　　　　　**B.** endothermic.

4. If the total energy change for the bond breaking steps is greater than for the bond making steps, the overall reaction will be

    A. exothermic　　　　　　**B.** endothermic.

    In questions 5 to 9 decide whether the enthalpy change in each of the processes is

    A. endothermic　　　　　　**B.** exothermic.

5. $Cl_2 (g) \rightarrow 2Cl (g)$

6. $Na (s) \rightarrow Na^+ (g)$

7. $Na (g) \rightarrow Na^+ (g) + e^-$

8. $Na^+ (g) + Cl^- (g) \rightarrow Na^+Cl^- (s)$

9. $Cl (g) + e^- \rightarrow Cl^- (g)$

10. For exothermic reactions, the sign for the enthalpy change has a

    A. a negative value　　　　**B.** a positive value.

11. For endothermic reactions, the sign for the enthalpy change has a

    A. a positive value　　　　**B.** a negative value.

12. The enthalpy change for a reaction is the energy difference between which of the following?

    A.   reactants and activated complex
    B.   products and activated complex
    C.   reactants and products

13. Which of the following is **not** a factor that affects the rate of a reaction?

    A.   concentration of reactants
    B.   kinetic energies of reactants
    C.   activation energy of reaction
    D.   enthalpy change for the reaction

Questions 14 to 17 refer to the effect of a catalyst.

    In each of the questions decide whether the statement is

    A.   TRUE   B.   FALSE.

14. It increases the enthalpy change.

15. It increases the energy of activation.

16. It decreases the enthalpy change.

17. It decreases the energy of activation.

18. A potential energy diagram can be used to show the activation energy (Ea) and the enthalpy change (ΔH) for a reaction.

    Which of the following combinations of Ea and ΔH could never be obtained for a reaction?

    A.   Ea = 50 kJ mol^{-1} and ΔH = -100 kJ mol^{-1}
    B.   Ea = 50 kJ mol^{-1} and ΔH = +100 kJ mol^{-1}
    C.   Ea = 100 kJ mol^{-1} and ΔH = +50 kJ mol^{-1}
    D.   Ea = 100 kJ mol^{-1} and ΔH = -50 kJ mol^{-1}

# Test 1.3	Reaction profiles

Questions 1 to 3 refer to the potential energy diagram.

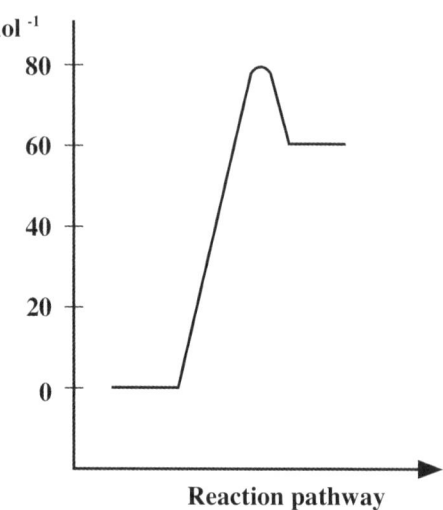

1. What is the enthalpy change, in kJ mol^{-1}, for the forward reaction?

   A. -60	B. -20
   C. +60	D. +80

2. What is the activation energy, in kJ mol^{-1}, for the forward reaction?

   A. 20	B. 60
   C. 40	D. 80

3. What is the activation energy, in kJ mol^{-1}, for the reverse reaction?

   A. 20	B. 60
   C. 40	D. 80

Questions 4 to 7 refer to the potential energy diagram.

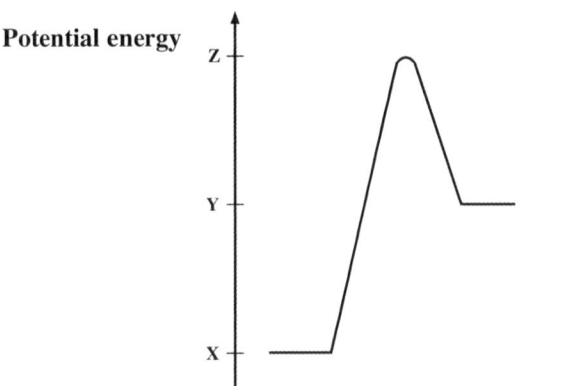

4. The activation energy for the forward reaction is given by

   A. Y   B. Z - X
   C. Y - X   D. Z.

5. The energy of activation for the reverse reaction is given by

   A. Y   B. Z - Y
   C. Y - X   D. Z.

6. The enthalpy change for the forward reaction is given by

   A. Z - Y   B. Y - Z
   C. Y - X   D. X - Y.

7. The enthalpy change for the reverse reaction is given by

   A. Z - Y   B. Y - Z
   C. Y - X   D. X - Y.

Chemical Changes and Structure

Test 1.5

Forms of carbon

Questions 1 and 2 refer to types of bonding and structure.

    A. covalent network
    B. covalent molecular
    C. monatomic

What is the type of bonding and structure in each of the following forms of carbon?

1. diamond

2. fullerenes

In questions 3 to 9 decide whether each of the statements refers to

    A. diamond   **B.** graphite.

3. has delocalised electrons

4. has layers held together by weak intermolecular forces

5. has a three-dimensional structure based on tetrahedrons

6. is used as a cutting tip

7. is used as a lubricant

8. is used as 'lead' in pencils

9. is a conductor of electricity

10. Which element has a similar crystal structure to diamond?

    A. berylium
    B. boron
    C. phosphorus
    D. silicon

Test 1.6   Trends in the Periodic Table (i)

Questions 1 to 9 refer to trends associated with increasing atomic number in the alkali metals.

    Decide whether each of the statements is

    A.   TRUE   **B.**   FALSE.

1. The metallic bond strength increases.
2. The first ionisation energy decreases.
3. The covalent radius decreases.
4. The number of occupied energy levels decreases.
5. The electronegativity decreases.
6. The melting point increases.
7. The tendency to form positive ions increases.
8. The nuclear charge increases.
9. The relative atomic mass decreases.

Questions 10 to 18 refer to trends associated with increasing atomic number in the halogens.

    Decide whether each of the statements is

    A.   TRUE   **B.**   FALSE.

10. The reactivity decreases.
11. The density decreases.
12. The first ionisation energy increases.
13. The boiling point increases.
14. The nuclear charge increases.
15. The electronegativity increases.
16. The relative atomic mass increases.
17. The covalent radius increases.
18. The number of occupied energy levels increases.

Chemical Changes and Structure

Questions 19 to 23 refer to trends associated with increasing atomic number in the elements in the second period of the Periodic Table.

Decide whether each of the statements is

    A.   TRUE   **B.**   FALSE.

19. The first ionisation energy increases.
20. The number of occupied energy levels increases.
21. The nuclear charge decreases.
22. The covalent radius increases.
23. The electronegativity decreases.

Questions 24 to 28 refer to the element francium (atomic number 87).

Decide whether each of the statements is likely to be

    A.   TRUE   **B.**   FALSE.

24. It will resist corrosion.
25. It will form a covalent chloride.
26. It will form a soluble hydroxide.
27. It will be very reactive.
28. It will conduct electricity.

Questions 29 to 33 refer to the element astatine (atomic number 85).

Decide whether each of the statements is likely to be

    A.   TRUE   **B.**   FALSE.

29. It will exist as diatomic molecules.
30. It will be a gas at room temperature.
31. It will form an ionic compound with sodium.
32. It will conduct electricity.
33. It will form a covalent compound with hydrogen.

Test 1.7   Trends in the Periodic Table (ii)

In questions 1 to 10, decide whether each of the statements is

    A.   TRUE   **B.**   FALSE.

1. A magnesium atom is smaller than a barium atom.

2. A nitrogen atom is smaller than an oxygen atom.

3. A bromide ion is larger than a fluoride atom.

4. A magnesium ion is larger than a sulphide ion.

5. A sodium ion is smaller than a potassium ion.

6. A rubidium ion is smaller than a bromide ion.

7. A potassium ion is larger than a potassium atom.

8. A chloride ion is larger than a chlorine atom.

9. A fluoride ion is smaller than an oxide ion.

10. An iron(III) ion is smaller than an iron(II) ion.

Questions 11 and 16 refer to elements.

    A.   caesium   **B.**   lithium
    C.   fluorine   **D.**   iodine

11. Which element is made up of atoms with the smallest covalent radius?

12. Which element is made up of atoms with the largest covalent radius?

13. Which element has the highest electronegativity?

14. Which element has the lowest electronegativity?

15. Which element has the lowest first ionisation energy?

16. Which element has the highest first ionisation energy?

17. Which property of the Group 1 elements could be represented by the graph?

    A. the first ionisation energy
    B. the melting point
    C. the covalent radius
    D. the electronegativity

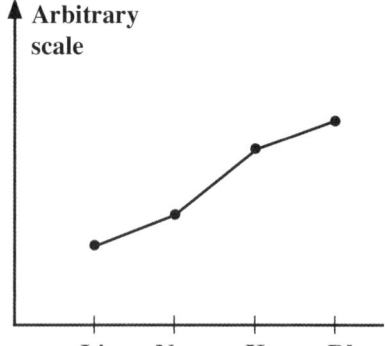

18. Potassium has a larger covalent radius than sodium because potassium has

    A. a larger nuclear charge
    B. a larger nucleus
    C. more occupied energy levels
    D. a larger first ionisation energy.

19. The difference between the covalent radius of sodium and chlorine is mainly due to the difference in the

    A. number of electrons
    B. number of protons
    C. number of neutrons
    D. mass of each atom.

Test 1.8   Ionisation energy

Questions 1 to 5 refer to what happens when an atom **X** of an element in Group 1 reacts to become an ion **X⁺**.

Decide whether each of the statements is

    A.   TRUE   **B.**   FALSE.

1. The covalent radius increases.

2. The nucleus acquires a negative charge.

3. The number of energy levels (electron shells) decreases by one.

4. The atomic number decreases by one.

5. An electron is emitted from the nucleus.

6. Which equation represents the first ionisation energy of calcium?

    A.   Ca (s)   →   Ca⁺ (g)   +   e⁻
    B.   Ca (s)   →   Ca⁺ (aq)   +   e⁻
    C.   Ca (g)   →   Ca⁺ (g)   +   e⁻
    D.   Ca (g)   →   Ca⁺ (aq)   +   e⁻

7. Which equation represents the first ionisation energy of fluorine?

    A.   F (g)   +   e⁻   →   F⁻ (g)
    B.   F (g)   →   F⁺ (g)   +   e⁻
    C.   ½ F₂ (g)   →   F⁻ (g)   +   e⁻
    D.   F⁺ (g)   +   e⁻   →   F (g)

8. Which equation represents the third ionisation energy of aluminium?

    A.   Al (s)   →   Al³⁺ (g)   +   3e⁻
    B.   Al (g)   →   Al³⁺ (g)   +   3e⁻
    C.   Al²⁺ (g)   →   Al³⁺ (g)   +   e⁻
    D.   Al³⁺ (g)   →   Al⁴⁺ (g)   +   e⁻

9. Which element would require the most energy to convert one mole of gaseous atoms into gaseous ions with a two-positive charge?

   Use the Ionisation Energies in the Data Booklet.

   A. scandium
   B. titanium
   C. vanadium
   D. chromium

10. The spike graph shows the variation in the first ionisation energy with atomic number for sixteen consecutive elements in the Periodic Table. The element at which the spike graph starts is not specified.

    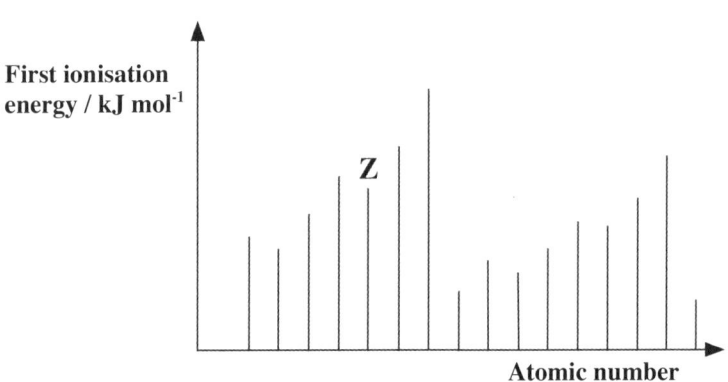

    In which group of the Periodic Table is element **Z**?

    A. 1   B. 3   C. 5   D. 6

Questions 11 and 12 refer to the ionisation energies of four elements

	1st Ionisation energy / kJ mol^{-1}	2nd Ionisation energy / kJ mol^{-1}	3rd Ionisation energy / kJ mol^{-1}
A.	1681	3374	6050
B.	496	4562	6910
C.	590	1145	4912
D.	578	1817	2745

11. Which element is most likely to form an ion of the type X^+?

12. Which element is most likely to form an ion of the type X^{2+}?

Chemical Changes and Structure

In questions 13 and 14, calculate the enthalpy change, in kJ mol^{-1}, for each of the reactions.

Use the Ionisation Energies in the Data Booklet.

13. K (g) → K^{2+} (g) + 2e$^-$

    A. +419
    B. +2633
    C. +3052
    D. +3471

14. Al^{3+} (g) + 2e$^-$ → Al$^+$ (g)

    A. +2167
    B. -2167
    C. +4562
    D. -4562

Test 1.9   Bonding and structure (ii)

Questions 1 to 8 refer to types of structure.

    A. covalent (molecular)   B. covalent (network)
    C. ionic

What type of structure predominates in each of the following compounds?

1. hydrogen bromide
2. sodium nitrate
3. silicon carbide
4. carbon tetrachloride
5. magnesium oxide
6. silicon dioxide
7. sulphur dioxide
8. potassium sulphate
9. phosphorus pentoxide
10. silicon tetrabromide

Questions 11 to 13 refer to the formation of chlorides.

    A. lithium   B. caesium
    C. sulphur   D. phosphorus
    E. xenon

11. Which element forms the chloride which is most ionic in character?
12. Which element forms the chloride which is most covalent in character?
13. Which element is least likely to form a chloride?

14. Which pair of elements would be expected to react together with the greatest release of energy?

    A. potassium and bromine
    B. caesium and fluorine
    C. lithium and iodine
    D. sodium and chlorine

15. Which statement may be applied to silicon dioxide?

    A. It consists of discrete molecules.
    B. It has a covalent network structure.
    C. It is similar in structure to carbon dioxide.
    D. It is made up of ions.

16. In the diagram, each sphere represents a particle about the size of an atom and the sign indicates the charge on the particle.

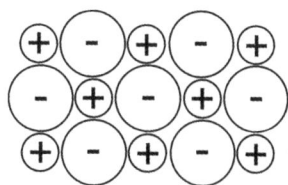

For which substance would the model be a reasonable representation of the particles and the way they are arranged in the crystal?

A. tetrabromomethane
B. calcium fluoride
C. lithium bromide
D. diamond

Questions 15 and 16 refer to the diagrams that represent molecules.

(i)    (ii)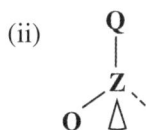

| represents an electron pair in the plane of the paper;

△ represents an electron pair in front of the plane of the paper;

╲ represents an electron pair behind the plane of the paper.

17. Which elements could make up structure (i) ?

	R	T
A.	nitrogen	hydrogen
B.	hydrogen	nitrogen
C.	carbon	hydrogen
D.	beryllium	chlorine

18. Which elements could make up structure (ii) ?

	Z	Q
A.	oxygen	hydrogen
B.	hydrogen	oxygen
C.	beryllium	hydrogen
D.	phosphorus	hydrogen

Chemical Changes and Structure

Test 1.10   Bonding and structure (iii)

Questions 1 to 7 refer to types of attraction.

    Decide whether each of the following

      A. is intermolecular   **B.** is **NOT** intermolecular.

1. covalent bonds
2. London dispersion forces
3. ionic bonds
4. permanent-permanent dipole attractions
5. hydrogen bonds
6. metallic bonds
7. van der Waals' forces

Questions 8 to 19 refer to van der Waals' forces of attraction.

    A. London dispersion forces
    B. hydrogen bonds
    C. permanent dipole-permanent dipole attractions but **no** hydrogen bonds

What is the main intermolecular attraction in each of the following substances?

8. neon
9. butane
10. methanol, CH_3-OH
11. hydrogen
12. hydrogen bromide
13. propanoic acid,
$$CH_3-\overset{\overset{O}{\|}}{C}-OH$$
14. hydrogen oxide
15. nitrogen hydride
16. ethyl ethanoate,
$$CH_3-\overset{\overset{O}{\|}}{C}-O-CH_2CH_3$$
17. hydrogen fluoride
18. octene
19. propanone,
$$CH_3-\overset{\overset{O}{\|}}{C}-CH_3$$

Chemical Changes and Structure

Test 1.11   Bonding and structure (iv)

Question 1 to 7 refer to types of attraction.

    A. non-polar covalent bonds   **B.** polar covalent bonds
    C. hydrogen bonds   **D.** London dispersion forces

What type of attraction is mainly responsible for holding each of the following pairs together?

1. two adjacent ethanol molecules

2. the carbon atom and a chlorine atom in a molecule of tetrachloromethene

3. two chlorine atoms in a molecule of chlorine

4. a hydrogen atom and the oxygen atom in a molecule of water

5. two adjacent hexene molecules in hexene

6. the carbon atom and an oxygen atom in a molecule of carbon dioxide

7. two adjacent molecules of hydrogen fluoride

Questions 8 to 11 refer to types of structure.

    A. three dimensional ionic lattice
    B. three dimensional covalently linked structure
    C. three dimensional structure of molecules, linked by hydrogen bonds
    D. linear covalent structure, linked by van der Waals' forces

Which type best describes the structure of each of the following?

8. hydrogen oxide

9. silicon dioxide

10. potassium chloride

11. polystyrene

Questions 12 to 21 refer to types of attraction.

    A. ionic bonds   **B.** hydrogen bonds
    C. London dispersion forces   **D.** covalent bonds
    E. permanent dipole-permanent dipole interactions

What is the main type of attraction that is overcome in converting each of the following from solid state to liquid state (melting)?

12. caesium fluoride
13. hydrogen fluoride
14. sulphur
15. silicon
16. ice
17. hydrogen sulphide
18. methane
19. boron
20. sulphur dioxide
21. rubidium chloride
22. hydrogen bromide
23. carbon monoxide

24. Which of the following shows the types of bonding forces in decreasing order of strength?

    A. covalent : hydrogen : London dispersion
    B. covalent : London dispersion: hydrogen
    C. hydrogen : covalent : London dispersion
    D. London dispersion: hydrogen : covalent

25. Which of the following occurs when crude oil is distilled?

    A. Covalent bonds break and form again.
    B. Covalent bonds break and van der Waals' bonds form.
    C. Van der Waals' bonds break and covalent bonds form.
    D. Van der Waals' bonds break and form again.

Test 1.12

Properties of elements and compounds (i)

Questions 1 to 6 refer to types of bonding and structure.

    A. covalent (discrete molecules)   **B.** ionic
    C. covalent (network structure)   **D.** metallic

Which type predominates in each of the following?

1. a substance which melts at 1044 °C and which conducts electricity when molten, but not when solid

2. a substance which melts at 962 °C and conducts electricity when solid

3. a substance which melts at 843 °C and boils at 1540 °C; when an electric current is passed through the molten substance no decomposition occurs

4. a substance of melting point 2300 °C, boiling point of 2550 °C, which does not conduct electricity

5. a substance melting at 1074 °C and boiling at 1740 °C; the passage of an electric current through the molten substance results in decomposition

6. a substance which melts at 6 °C and boils at 80 °C, which does not conduct electricity

Questions 7 to 10 refer to properties of elements and compounds.

    A. potassium fluoride   **B.** silicon oxide
    C. sulphur   **D.** sodium

7. Which is a solid of low melting point with high electrical conductivity?

8. Which is a non-conducting solid which becomes a good conductor on melting?

9. Which is a solid of high melting point with no electrical conductivity?

10. Which is a solid of low melting point with no electrical conductivity?

11. Information about four solids, **A**, **B**, **C** and **D**, is shown.

    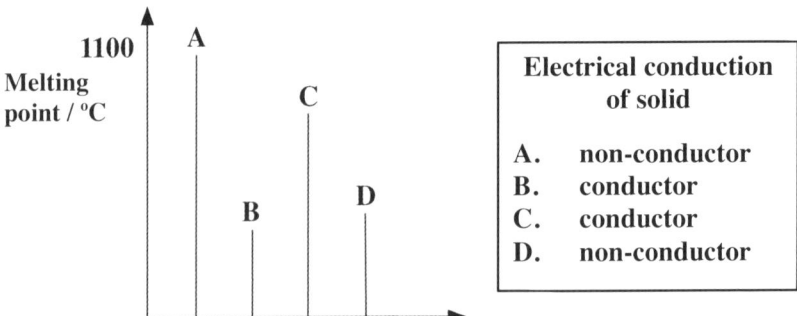

    In which solid is it most likely that only London dispersion forces are overcome when the substance melts?

12. Molten lithium hydride can be electrolysed using platinum electrodes.

    What is the reaction taking place at the positive electrode?

    A. $2H^- (l) \rightarrow H_2 (g) + 2e^-$
    B. $2H^+ (l) + 2e^- \rightarrow H_2 (g)$
    C. $Li^+ (l) + e^- \rightarrow Li (l)$
    D. $Li (l) \rightarrow Li^+ (l) + e^-$

13. Which property of a chloride would prove that it contained ionic bonding?

    A. It conducts electricity when molten.
    B. It is soluble in a polar solvent.
    C. It is a solid at room temperature.
    D. It has a high boiling point.

Test 1.13

Properties of elements and compounds (ii)

In questions 1 to 8 decide whether each of the compounds, at the temperature given, is

- **A.** a solid
- **B.** a liquid
- **C.** a gas.

Use the Melting and Boiling Points in the Data Booklet.

1. magnesium oxide (MgO) at 3000 °C

2. beryllium chloride (BeCl$_2$) at room temperature

3. dinitrogen tetroxide (N$_2$O$_4$) at 0 °C

4. boron oxide (B$_2$O$_3$) at 500 °C

5. fluorine chloride (FCl) at -50 °C

6. phosphorus chloride (PCl$_3$) at -10 °C

7. sulphur dioxide (SO$_2$) at room temperature

8. silicon chloride (SiCl$_4$) at -80 °C

Questions 9 and 10 refer to types of bonding.

- **A.** metallic
- **B.** polar covalent
- **C.** non-polar covalent
- **D.** ionic

Which type of bonding is likely to be present in each of the following?

9. an element which melts at 3500 °C and forms a gaseous oxide

10. a compound of nitrogen which boils at -33 °C

11. In general, covalent substances have lower melting points than ionic substances because

    A. ionic bonds are stronger than covalent bonds
    B. covalent compounds are composed of non-metals which have low melting points
    C. bonds between molecules are weaker than bonds between ions
    D. covalent bonds have no electrostatic forces.

12. Carbon dioxide is a gas at room temperature while silicon dioxide is a solid because

    A. London dispersion forces are much weaker than covalent bonds
    B. carbon dioxide contains double covalent bonds and silicon dioxide contains single covalent bonds
    C. carbon-oxygen bonds are less polar than silicon-oxygen bonds
    D. the relative formula mass of carbon dioxide is less than that of silicon dioxide.

13. The diagram shows the melting points of successive elements across a period in the Periodic Table.

    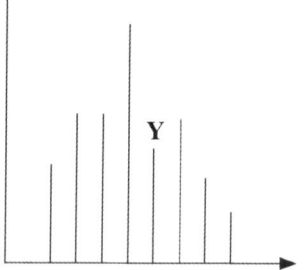

    Which is a correct reason for the low melting point of element **Y**?

    A. It has weak metallic bonds.
    B. It has weak covalent bonds.
    C. It has weakly held outer electrons.
    D. It has weak forces between molecules.

14.
                OH OH   Cl H
                | | | |
        H−C−C−H   H−C−C−H
              | | | |
              H   H   H   H

        Substance **X**   Substance **Y**
        Molecular mass 62   Molecular mass 64.5

From a consideration of chemical bonding, what can you predict about the boiling points of these compounds?

    A.   The boiling point of **X** is greater than that of **Y**.
    B.   The boiling point of **X** is less than that of **Y**.
    C.   The boiling point of **X** is approximately equal to that of **Y**.
    D.   Nothing can be predicted.

15. Which chloride is most likely to be soluble in tetrachloromethane?

    A.   barium chloride
    B.   caesium chloride
    C.   calcium chloride
    D.   phosphorus chloride

16. Which substance is insoluble in water but soluble in tetrachloromethane?

    A.   iodine   **B.**   sodium chloride
    C.   potassium iodide   **D.**   lithium bromide

17. An ionic compound is likely to

    A.   have a low melting point
    B.   dissolve in non-polar solvents
    C.   be an electrical insulator when molten
    D.   be soluble in water.

18. Silicon carbide can be used as

    A.   a lubricant
    B.   a tip for cutting / grinding tools
    C.   a substitute for pencil 'lead'
    D.   an electrical conductor.

Test 1.14 Polarity of molecules

In questions 1 to 10 decide whether each of the molecules

- **A.** has polar bonds
- **B.** has non-polar bonds
- **C.** is overall polar
- **D.** is overall non-polar
- **E.** is three-dimensional
- **F.** is planar
- **G.** is linear.

For each of the questions, **three** answers should be given.

You may wish to use the Data Booklet.

1. H_2
2. CCl_4
3. HCl
4. H_2O
5. CO_2
6. NH_3
7. PH_3
8. $CHCl_3$
9. CH_3OH
10. CS_2

11. In which molecule will the chlorine atom carry a partial positive charge (δ+)?

    A. Cl - Br **B.** Cl - Cl **C.** Cl - F **D.** Cl - I

12. Nitrogen will form a non-polar covalent bond with an element with an electronegativity of

    A. 1.5 **B.** 2 **C.** 2.5 **D.** 3.0.

Chemical Changes and Structure

Test 1.15 **Problem solving**

1. 1 mol of hydrogen gas reacts with 1 mol of iodine vapour.
   After **t** seconds, 0.8 mol of hydrogen remains.

   What is the number of moles of hydrogen iodide formed at **t** seconds?

   A. 0.2   B. 0.4   C. 0.8   D. 1.6

2. The results of an experiment carried out at 19 °C involving the reaction between equal volumes of 0.5 mol l^{-1} nitric acid and sodium thiosulphate solution of different concentrations are shown.

Concentration of sodium thiosulphate solution / mol l^{-1}	0.5	0.25	0.125	0.064
Time for the appearance of sulphur / s	13	25	51	104

   On the evidence of these results alone, which statement is correct?

   A. The more concentrated the thiosulphate solution, the longer the time before the sulphur appears.
   B. The more concentrated the nitric acid, the faster the reaction proceeds.
   C. The more concentrated the thiosulphate solution, the faster the reaction proceeds.
   D. The higher the temperature, the faster the reaction proceeds.

3. The continuous use of large extractor fans greatly reduces the possibility of an explosion in a flour mill.

   This is mainly because

   A. a build-up in the concentration of oxygen is prevented
   B. local temperature rises are prevented by the movement of air
   C. particles of flour suspended in the air are removed
   D. the slow accumulation of carbon monoxide is prevented.

Chemical Changes and Structure

4. Two identical samples of zinc are placed in open beakers. Excess of 2 mol l⁻¹ sulphuric acid is added to one, and excess of 1 mol l⁻¹ sulphuric acid is added to the other. All other conditions are the same.

   Which of the following is the same for the two samples?

   A. the mass lost from the beakers
   B. the total time for the reaction
   C. the initial reaction rate
   D. the average rate of evolution of gas

5. Magnesium was added to an excess of dilute hydrochloric acid, concentration 1 mol l⁻¹.

   Which measurement taken at regular intervals and plotted against time, would give the graph shown?

   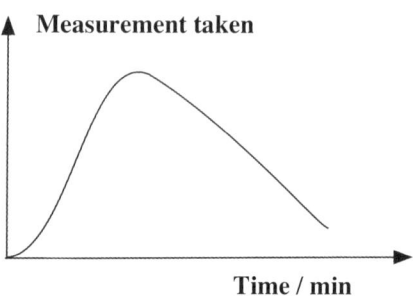

   A. the temperature of the solution
   B. the volume of hydrogen produced
   C. the pH of the solution
   D. the concentration of the solution

6. Excess marble chips (calcium carbonate) were added to 25 cm³ of hydrochloric acid, concentration 2 mol l⁻¹.

   Which measurement taken at regular intervals and plotted against time, would give the graph shown?

   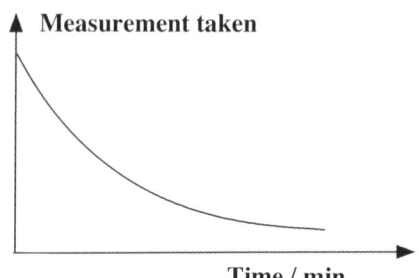

   A. temperature
   B. volume of gas produced
   C. pH of the solution
   D. mass of the beaker and contents

7. Excess calcium carbonate was added to 100 cm³ of 1 mol l⁻¹ hydrochloric acid. The experiment was repeated using the same mass of marble chips and 100 cm³ of 1 mol l⁻¹ sulphuric acid.

   Which would have been the same for both experiments?

   A. the time taken for the reaction to be completed
   B. the rate at which the first 10 cm³ of gas is eveolved
   C. the mass of marble chips left over when reaction has stopped
   D. the average rate of reaction

8. Which graph could apply to the following reaction?

   $S_2O_3^{2-}$ (aq) + 2H⁺ (aq) → H_2O (l) + SO_2 (g) + S (s)

   A.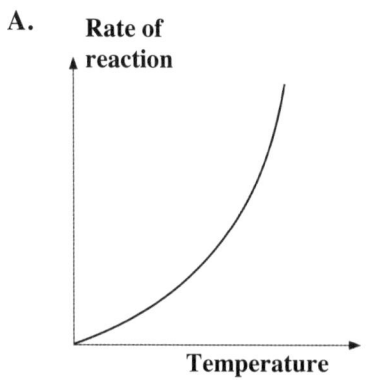

   B.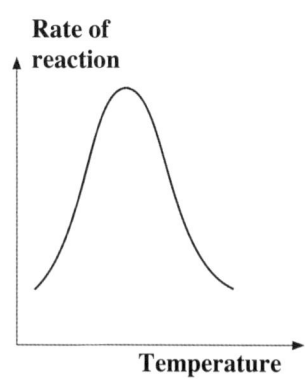

   C.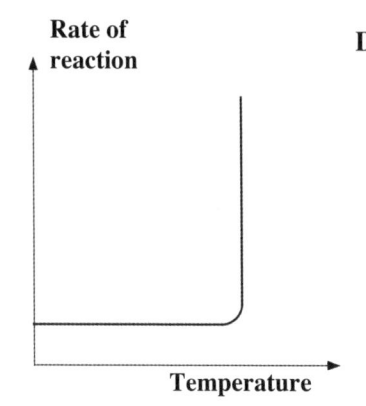

   D.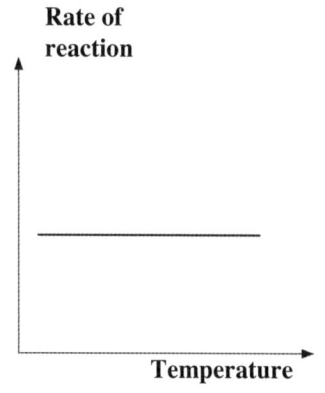

9. The graph shows the data obtained from three reactions of zinc with an excess of 2 mol l⁻¹ hydrochloric acid.

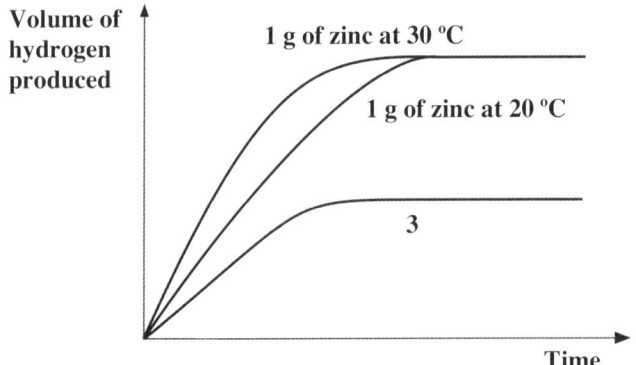

Which statement is true?

A. Increasing the temperature increases the total mass of hydrogen produced.
B. Increasing the temperature has no effect on the initial rate of reaction.
C. Curve 3 would be obtained with 1 g of zinc at 10 °C.
D. Curve 3 would be obtained with 0.5 g of zinc at 20 °C.

Questions 10 and 11 refer to the graph which shows the reactions of three metals with excess of 2 mol l⁻¹ hydrochloric acid.

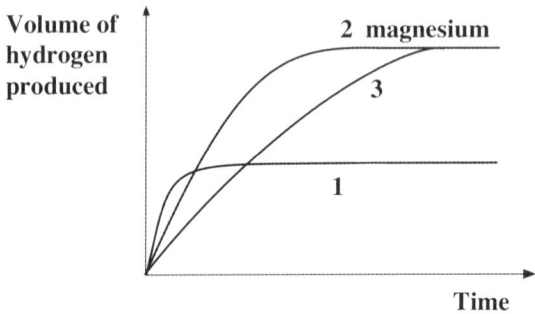

10. Which metal is likely to produce curve 1?

    A. lithium   B. aluminium   C. zinc   D. copper

11. Which metal is likely to produce curve 3?

    A. lithium   B. aluminium   C. zinc   D. copper

Questions 12 to 17 refer to the graphs which show data obtained from reactions of hydrochloric acid.

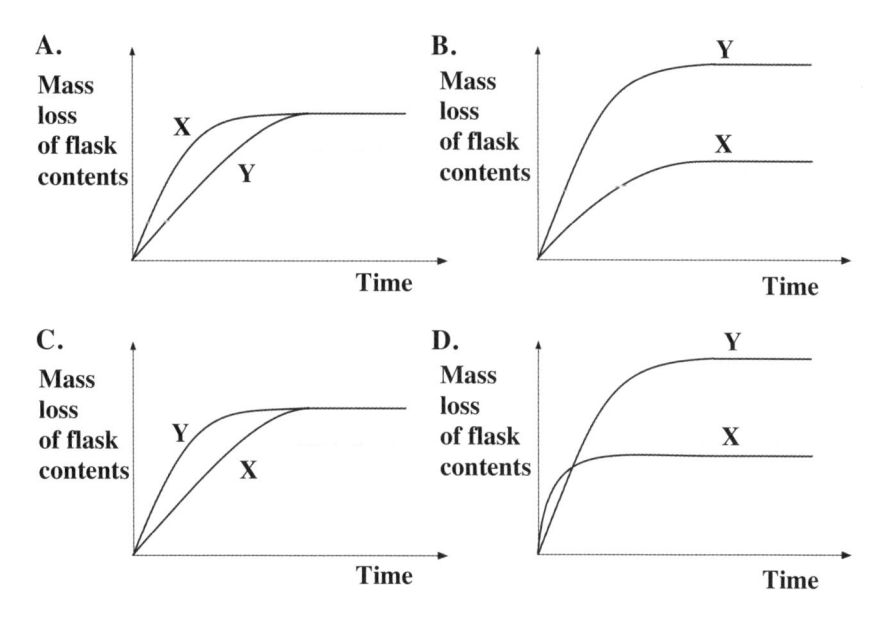

A.
Mass loss of flask contents vs Time — curve X rises faster than curve Y to the same plateau.

B.
Mass loss of flask contents vs Time — curve Y plateaus higher than curve X.

C.
Mass loss of flask contents vs Time — curve Y rises faster than curve X to the same plateau.

D.
Mass loss of flask contents vs Time — curve Y plateaus higher than curve X.

Which graph shows the data likely to be obtained from each of the following pairs of reactions?

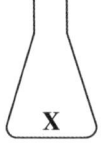

	Contents of flask **X**	Contents of flask **Y**
12.	10 g chalk lumps (excess) 50 cm³ of 1 mol l⁻¹ HCl (aq) 20 °C	10 g chalk powder (excess) 50 cm³ of 1 mol l⁻¹ HCl (aq) 20 °C
13.	4 cm magnesium ribbon 50 cm³ of 2 mol l⁻¹ HCl (aq) (excess) 20 °C	4 cm magnesium ribbon 50 cm³ of 1 mol l⁻¹ HCl (aq) (excess) 20 °C

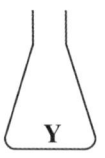

Chemical Changes and Structure

14. 10 g chalk (excess)
    50 cm³ of 0.1 mol l⁻¹ HCl (aq)
    20 °C

    10 g chalk (excess)
    50 cm³ of 0.2 mol l⁻¹ HCl (aq)
    20 °C

15. 4 cm magnesium ribbon
    50 cm³ of 2 mol l⁻¹ HCl (aq)
    (excess)
    20 °C

    8 cm magnesium ribbon
    50 cm³ of 1 mol l⁻¹ HCl (aq)
    (excess)
    20 °C

16. 2 g zinc (excess)
    50 cm³ of 1 mol l⁻¹ HCl (aq)
    20 °C with catalyst

    2 g zinc (excess)
    50 cm³ of 1 mol l⁻¹ HCl (aq)
    20 °C without catalyst

17. 2 g zinc (excess)
    50 cm³ of 1 mol l⁻¹ HCl (aq)
    20 °C

    2 g zinc (excess)
    50 cm³ of 1 mol l⁻¹ HCl (aq)
    40 °C

18. When copper carbonate reacts with excess acid, carbon dioxide is produced.
    The curves shown were obtained under two different conditions.

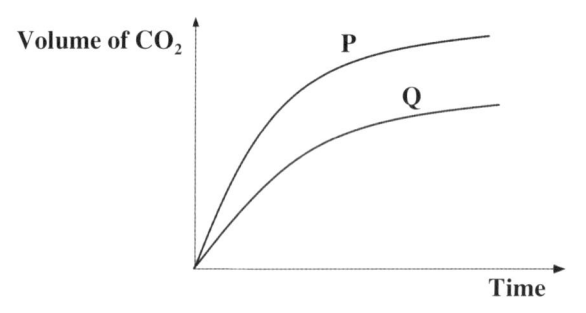

    The change from **P** to **Q** can be brought about by

    A. decreasing the concentration of the acid
    B. increasing the mass of copper carbonate
    C. decreasing the particle size of the copper carbonate
    D. increasing the temperature.

19. The course of the reaction between magnesite (magnesium carbonate) and dilute hydrochloric acid was followed by determining the mass of the reaction vessel and contents as carbon dioxide was evolved. The curves shown were obtained under two different conditions.

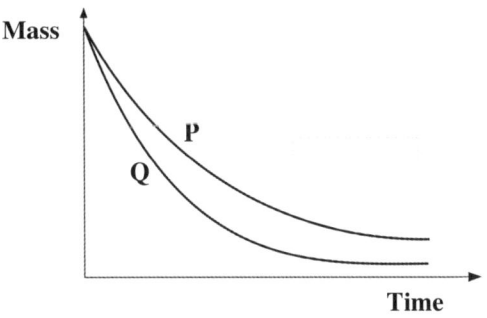

The change from **P** to **Q** can be brought about by

   A. decreasing the concentration of the acid
   B. increasing the temperature of the reactants
   C. increasing the particle size and mass of the magnesite
   D. decreasing the volume of the acid.

20. The graph shows how the rate of a reaction varies with the concentration of one of the reactants.

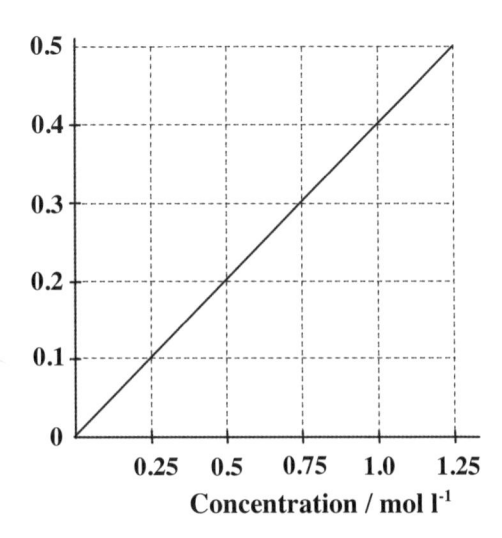

   What was the reaction time, in seconds, when the concentration of the reactant was 0.5 mol l^{-1}?

   A. 0.2   B. 0.5   C. 2.0   D. 5.0

21. The graph opposite shows the volume of hydrogen given off against time when an excess of magnesium ribbon is added to 100 cm³ of hydrochloric acid, concentration 1 mol l⁻¹, at 30 °C.

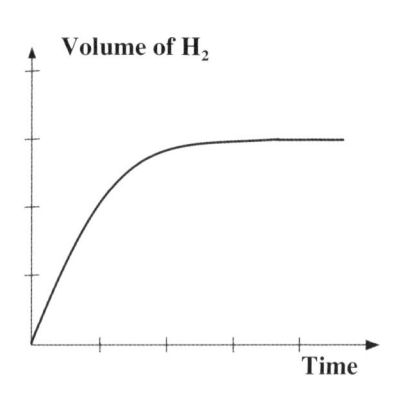

Which graph would show the volume of hydrogen given off when an excess of magnesium ribbon is added to 50 cm³ of hydrochloric acid of the same concentration at 20 °C?

A.

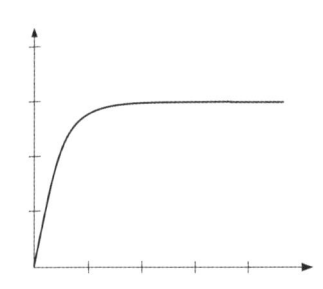

B.

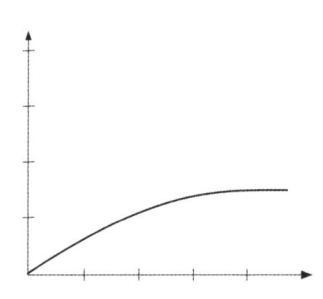

C.

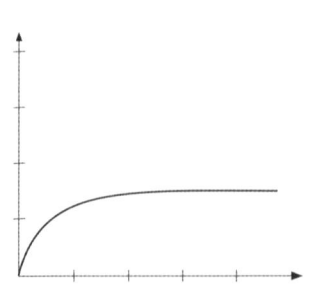

D.

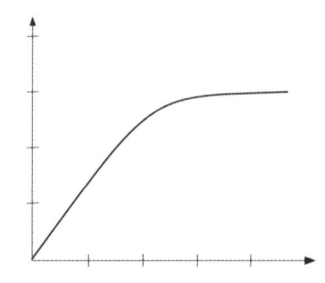

22. Excess zinc was added to 100 cm³ of hydrochloric acid, concentration 1 mol l⁻¹. Curve **1** refers to this reaction.

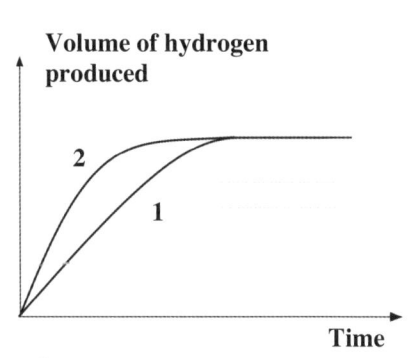

Curve **2** could be for

- **A.** excess zinc reacting with 100 cm³ of hydrochloric acid, concentration 2 mol l⁻¹
- **B.** excess zinc reacting with 100 cm³ of sulphuric acid, concentration 1 mol l⁻¹
- **C.** excess zinc reacting with 100 cm³ of nitric acid, concentration 1 mol l⁻¹
- **D.** excess magnesium reacting with 100 cm³ hydrochloric acid, concentration 1 mol l⁻¹.

23. Graph **X** was obtained when 1 g of calcium carbonate powder reacted with excess dilute hydrochloric acid.

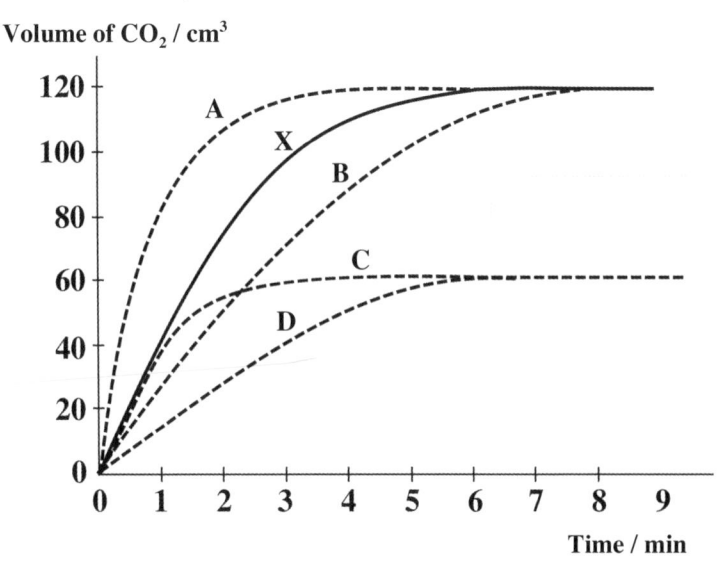

Which curve would best represent the reaction of 0.5 g lump calcium carbonate with excess of the same dilute hydrochloric acid?

Test 2.1 Structure of hydrocarbons (revision)

In questions 1 to 10 decide whether each of the hydrocarbons is

A. an alkane / cycloalkane **B.** an alkene / cycloalkene
C. saturated **D.** unsaturated.

For each of the questions, **two** answers should be given.

1. ethane

2. cyclopropane

3. propene

4. cycloheptene

5. octane

6. 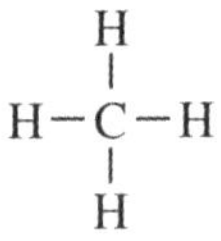

7. 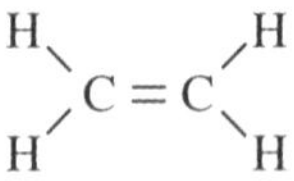

8. 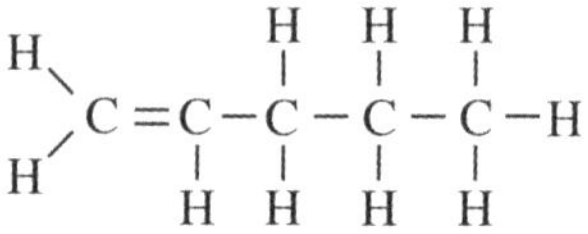

9. 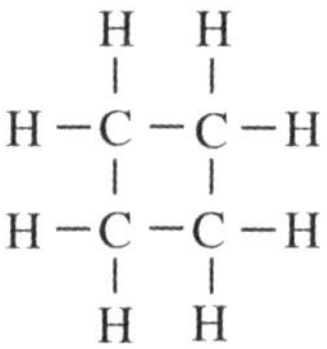

10. 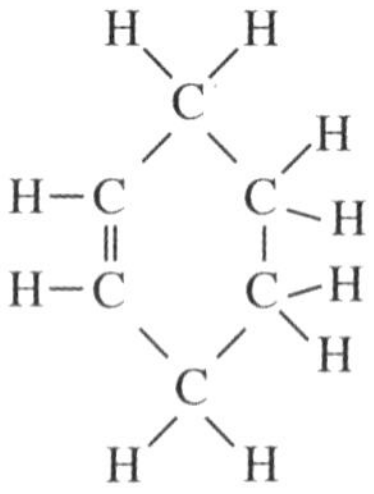

Questions 11 to 15 refer to the naming of hydrocarbons.

A.
$$H-\underset{\underset{H}{|}}{\overset{\overset{H}{|}}{C}}-H$$

B.
$$\underset{H}{\overset{H}{>}}C=C\underset{\overset{H}{}}{\overset{H}{<}}$$

C.
$$H-\underset{\underset{H}{|}}{\overset{\overset{H}{|}}{C}}-\underset{\underset{H}{|}}{\overset{\overset{H}{|}}{C}}-H$$

D.
$$\underset{H}{\overset{H}{>}}C=\underset{\underset{H}{|}}{\overset{\overset{H}{|}}{C}}-\underset{\underset{H}{|}}{\overset{\overset{H}{|}}{C}}-H$$

E.
$$H-\underset{\underset{H}{|}}{\overset{\overset{H}{|}}{C}}-\underset{\underset{H}{|}}{\overset{\overset{H}{|}}{C}}-\underset{\underset{H}{|}}{\overset{\overset{H}{|}}{C}}-\underset{\underset{H}{|}}{\overset{\overset{H}{|}}{C}}-H$$

F.
$$H-\underset{\underset{H}{|}}{\overset{\overset{H}{|}}{C}}-\underset{\underset{H}{|}}{\overset{\overset{H}{|}}{C}}-\underset{\underset{H}{|}}{\overset{\overset{H}{|}}{C}}-H$$

G.
$$H-\underset{\underset{H}{|}}{\overset{\overset{H}{|}}{C}}-\overset{\overset{H}{|}}{C}=\overset{\overset{H}{|}}{C}-\underset{\underset{H}{|}}{\overset{\overset{H}{|}}{C}}-H$$

H.
Cyclopropane structure (triangle of 3 carbons, each with 2 H)

I.
$$\underset{H}{\overset{H}{>}}C=\overset{\overset{H}{|}}{C}-\underset{\underset{H}{|}}{\overset{\overset{H}{|}}{C}}-\underset{\underset{H}{|}}{\overset{\overset{H}{|}}{C}}-H$$

J.
Cyclobutene structure (4-carbon ring with C=C double bond)

K.
Cyclobutane structure (4-carbon ring, each C with 2 H)

11. Which is methane?

12. Which is a but-1-ene?

13. Which is propene?

14. Which is cyclopropane?

15. Which is cyclobutene?

Questions 16 to 19 refer to the formulae of C_6 hydrocarbons.

    A. C_6H_{10} **B.** C_6H_{12} **C.** C_6H_{14}

What is the formula for each of the following hydrocarbons?

16. hexane

17. cyclohexane

18. hexene

19. cyclohexene

20. How many hydrogen atoms are in a straight-chain alkane with 25 carbon atoms?

    A.   48   **B.**   50   **C.**   52   **D.**   54

21. How many hydrogen atoms are in a straight-chain alkene with 12 carbon atoms?

    A.   20   **B.**   22   **C.**   24   **D.**   26

22. How many hydrogen atoms are in a cycloalkane with 16 carbon atoms?

    A.   28   **B.**   30   **C.**   32   **D.**   34

23. How many hydrogen atoms are in a cycloalkene with 21 carbon atoms?

    A.   40   **B.**   42   **C.**   44   **D.**   46

24. Which of the following could **not** be either a straight-chain alkane or a cycloalkane?

    A. C_4H_{10} **B.** C_5H_{10} **C.** C_6H_{12} **D.** C_7H_{12}

In questions 25 to 29 decide which hydrocarbon is **not** a member of the same homologous series as the others.

25. A. ethene   B. hexene   C. pent-1-ene   D. cyclopropane

26. A. butane   B. methane   C. octane   D. propene

27. A. C_3H_8 B. C_5H_{12} C. C_6H_{12} D. C_7H_{16}

28. A. CH_4 B. C_3H_6 C. C_6H_{12} D. C_8H_{16}

29. Hydrocarbons with a formula mass of :

    A. 16   B. 44   C. 84   D. 100

Questions 30 and 31 refer to homologous series.

    A. CH_4 B. $CH_2=CH_2$ C. 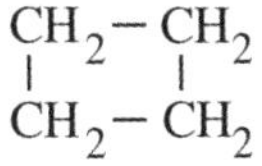 $\begin{array}{c} CH_2-CH_2 \\ | \quad\quad | \\ CH_2-CH_2 \end{array}$

Which hydrocarbon is a member of the same homologous series as each of the following?

30. $CH_3CH_2CH_3$

31. $CH_3CHCHCH_3$

Test 2.2 Systematic naming of hydrocarbons (revision)

Questions 1 to 10 refer to the systematic naming of hydrocarbons.

A.	2,3-dimethylbutane	B.	2-ethylbutane
C.	2,2-dimethylbutane	D.	2-methylbut-1-ene
E.	3-methylbut-1-ene	F.	1,1-dimethylcyclobutane
G.	3-methylpentane	H.	2-methylpentane
I.	pent-2-ene	J.	pent-1-ene
K.	4-methylpent-2-ene	L.	3,4-dimethylpent-2-ene
M.	2-methylpent-2-ene	N.	2,3-dimethylpent-2-ene
O.	4-ethyl-5-methylhex-2-ene	P.	4,4-dimethylhex-2-ene
Q.	2-methycyclohexene	R.	3-methylcyclohexene

What is the systematic name for each of the following hydrocarbons?

1. $CH_3-\underset{\underset{H}{|}}{\overset{\overset{CH_3}{|}}{C}}-\underset{\underset{H}{|}}{\overset{\overset{CH_3}{|}}{C}}-CH_3$

2. $CH_3-\underset{\underset{\underset{CH_3}{|}}{CH_2}}{\overset{\overset{H}{|}}{C}}-CH_2-CH_3$

3. $CH_3-C=C-\underset{|}{\overset{\overset{CH_3}{|}}{C}}-CH_3$ with H, H, H below

4. $CH_3-CH_2-CH_2-\underset{\underset{H}{|}}{\overset{\overset{CH_3}{|}}{C}}-CH_3$

5. $\underset{H}{\overset{H}{\diagdown}}C=C-\underset{\underset{H}{|}}{\overset{\overset{CH_3}{|}}{C}}-CH_3$ with H below second C

6. $CH_3-C=C-\underset{\underset{CH_3}{|}}{\overset{\overset{H}{|}}{C}}-CH_3$ with CH_3, H above/below

7. $\underset{\underset{CH_3}{|}}{\overset{\overset{CH_2-CH_2}{|\quad|}}{CH_2-C}}-CH_3$

8. $CH_3-CH_2-CH_2-\underset{|}{\overset{|}{C}}=C\underset{H}{\overset{H}{\diagup}}$ with H below

9. $CH_3-C=C-\underset{\underset{\underset{CH_3}{|}}{H-C-CH_3}}{C}-CH_2-CH_3$ with H H H

10. cyclohexene ring structure with CH_3 substituent

46   Nature's Chemistry

Questions 11 to 18 refer to the systematic naming of hydrocarbons.
Each structural formula is written without showing bonds and using brackets to show the position of a branch,

e.g.
$$CH_3-\underset{\underset{CH_3}{|}}{\overset{\overset{H}{|}}{C}}-CH_2-CH_3$$
is written: $CH_3CH(CH_3)CH_2CH_3$

A.	methylpropane	B.	propene
C.	methylpropene	D.	2,3-dimethylbutane
E.	but-1-ene	F.	methylbut-2-ene
G.	3,3-dimethybut-1-ene	H.	3-ethylpentane
I.	3,3-dimethylpentane	J.	pent-2-ene
K.	3-methylhex-2-ene	L.	4-methylhex-2-ene

What is the systematic name for each of the following hydrocarbons?

11. $CH_3CH(CH_3)CH(CH_3)CH_3$

12. $(CH_3)_3CH$

13. $CH_3CHC(CH_3)_2$

14. CH_2CHCH_3

15. $CH_3CH_2C(CH_3)_2CH_2CH_3$

16. $(CH_3)_2CCH_2$

17. $CH_3C(CH_3)_2CHCH_2$

18. $CH_3CHCHCH(CH_3)CH_2CH_3$

19. Which hydrocarbon has a molecular formula different from that of the other three?

   A. tetramethylbutane
   B. 3-ethylpentane
   C. 2,3-dimethylpentane
   D. trimethylbutane

20. Which hydrocarbon has the same formula mass as 3,3-dimethylbut-1-ene?

   A. 2,2-dimethylbutane
   B. 2-methylbut-1-ene
   C. 2-methylpent-1-ene
   D. 3-methylcyclopentene

Test 2.3 Isomeric hydrocarbons (revision)

Decide whether each of the pairs of compounds

A. are isomers **B.** are **NOT** isomers.

1. $CH_3-CH_2-CH_2-CH_3$ and $CH_3-CH_2-CH_2-CH_3$ (branched: CH_3-CH_2 with CH_2-CH_3)

2. $CH_3-CH(CH_3)-CH_3$ and $CH_3-CH_2-CH_2-CH_3$

3. $CH_3-CH(CH_3)-CH_3$ and $CH_3-CH(CH_3)-CH_3$

4. $CH_3-CH_2-CH_3$ and $CH_2=CH-CH_3$

5. $CH_3-CH_2-CH_3$ and CH_3-CH_3

6. $CH_3-CH_2-CH_2-CH_2-CH_3$ and $CH_3-CH(CH_3)-CH_2-CH_3$

7. $CH_3-CH(CH_3)-CH_2-CH_3$ and $CH_3-CH_2-CH(CH_3)-CH_3$

8. $CH_3-CH_2-CH_2-CH_2-CH_3$ and $CH_3-C(CH_3)(CH_3)-CH_3$

9. $CH_3-CH(CH_3)-CH_2-CH_3$ and $CH_3-C(CH_3)(CH_3)-CH_3$

10. 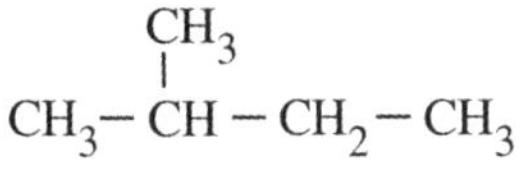 $CH_3-\underset{\underset{CH_3}{|}}{CH}-CH_2-CH_3$ and $CH_3-CH_2-\underset{\underset{CH_3}{|}}{CH}-CH_3$

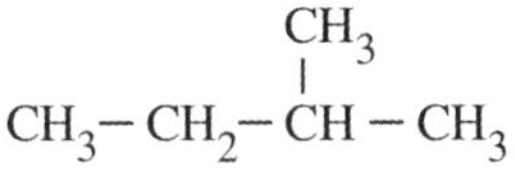

11. $CH_3-CH_2-CH_2-CH_2-CH_3$ and $\underset{\underset{CH_3}{|}}{CH_2}-CH_2-\underset{\underset{CH_3}{|}}{CH_2}$

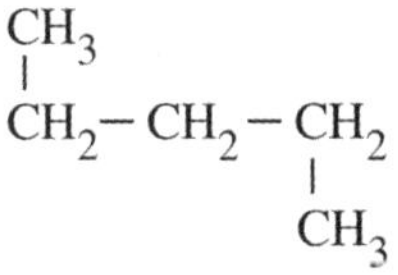

12. $CH_2=CH-CH_2-CH_3$ and $CH_3-CH_2-CH=CH_2$

13. $CH_2=CH-CH_2-CH_3$ and $CH_3-CH=CH-CH_3$

14. $CH_2=CH-CH_2-CH_3$ and $\begin{array}{c}CH_2-CH_2\\||\\CH_2-CH_2\end{array}$

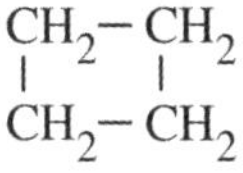

15. $CH_3-CH_2-CH_2-CH_3$ and $\begin{array}{c}CH_2-CH_2\\||\\CH_2-CH_2\end{array}$

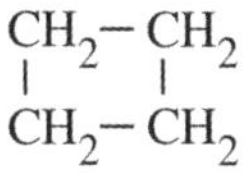

16. $CH_3-CH_2-CH_2-CH_3$ and $CH_3-\underset{\underset{CH_3}{|}}{C}=CH-CH_3$

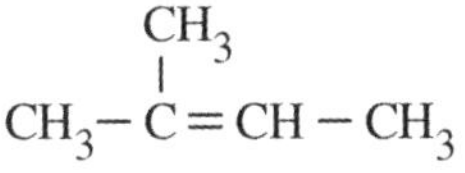

17. 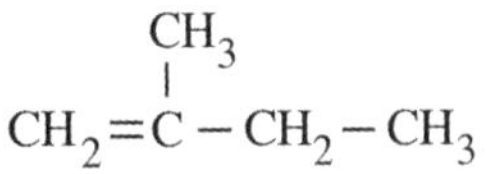 $CH_2=\underset{\underset{CH_3}{|}}{C}-CH_2-CH_3$ and $CH_3-\underset{\underset{CH_3}{|}}{C}=CH-CH_3$

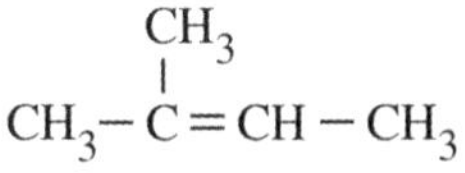

18. 2-methylpentane and heptane

19. 2,3-dimethylbutane and hexane

20. 2-methylpent-1-ene and hex-1-ene

21. methylpropane and butane

22. but-2-ene and 2-methylbut-1-ene

Nature's Chemistry

Test 2.4   Functional groups (i)

The questions in this test refer to oxygen-containing organic compounds.

A. butanone   **B.** pentanal   **C.** methanol

D. ethyl methanoate   **E.** CH_3-O-CH_3 **F.** $H-C\underset{OH}{\overset{O}{\lessgtr}}$

G. $CH_3-\underset{\|}{\overset{O}{C}}-CH_3$ **H.** $CH_3-\underset{\|}{\overset{O}{C}}-O-CH_2CH_3$

I. $H-\underset{\|}{\overset{O}{C}}-H$ **J.** $CH_3-CH_2-\underset{\|}{\overset{O}{C}}-H$

K. $CH_3-O-\underset{\|}{\overset{O}{C}}-H$ **L.** $CH_3-O-CH_2-CH_3$

M. $CH_3-\underset{\underset{H}{|}}{\overset{\overset{H}{|}}{C}}-\underset{\underset{OH}{|}}{\overset{\overset{CH_3}{|}}{C}}-CH_3$ **N.** $H-\underset{\|}{\overset{O}{C}}-CH_3$ **O.** cyclohexanone structure

P. CH_3COOH **Q.** CH_3OH

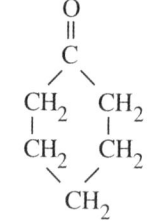

R. $CH_3COCH_2CH_3$ **S.** $CH_3CH(OH)CH_3$

T. $HCOOCH_2CH_3$ **U.** $CH_3CH_2OCOCH_3$

V. $CH_3CH_2CH_2CHO$ **W.** $CH_3CH_2OCH_2CH_2CH_3$

1. Pick out **all** the compounds that are alcohols.
2. Pick out **all** the compounds that are organic acids.
3. Pick out **all** the compounds that are aldehydes.
4. Pick out **all** the compounds that are ketones.
5. Pick out **all** the compounds that are esters.

Nature's Chemistry

Test 2.5 Simple esters – structures and names

Questions 1 to 4 refer to names of esters.

1. What ester is formed in the reaction of methanol with ethanoic acid?

   A. methyl ethanoate   **B.** ethyl methanoate

2. What ester is formed in the reaction of propanol with methanoic acid?

   A. methyl propanoate   **B.** propyl methanoate

3. What ester is formed in the reaction of ethanoic acid with butanol?

   A. butyl ethanoate   **B.** ethyl butanoate

4. What ester is formed in the reaction of butanoic acid with methanol?

   A. butyl methanoate   **B.** methyl butanoate

Questions 5 to 8 refer to structures of esters.

A. $CH_3-\overset{\overset{O}{\|}}{C}-O-CH_3$ **B.** $CH_3CH_2-\overset{\overset{O}{\|}}{C}-O-CH_3$

C. $CH_3CH_2-O-\overset{\overset{O}{\|}}{C}-CH_3$ **D.** $CH_3CH_2-O-\overset{\overset{O}{\|}}{C}-H$

5. What is the structure of the ester formed in the reaction between ethanol and methanoic acid?

6. What is the structure of the ester formed in the reaction between propanoic acid and methanol?

7. What is the structure of ethyl ethanoate?

8. What is the structure of methyl ethanoate?

9. The formulae for the molecules from which ethylbutanoate is made are

   A. C_3H_7OH and CH_3COOH
   B. C_2H_5OH and C_2H_5COOH
   C. C_3H_7COOH and C_2H_5OH
   D. C_2H_5COOH and C_3H_7OH.

Test 2.6 — Isomers again

In questions 1 to 6 decide whether each of the pairs of compounds

    A. are isomers   **B.** are NOT isomers.

1. Br–CH₂–CH₂–Br and H–CHBr–CHBr–H

2. Br–CH₂–CH₂–Br and H–CBr₂–CH₂–H

3. $CH_3-CH_2-CH_2-OH$ and $CH_3-CH(OH)-CH_3$

4. $CH_3-C(=O)-O-CH_3$ and $CH_3-CH_2-C(=O)-OH$

5. $CH_3-C(=O)-CH_3$ and $CH_3-CH_2-C(=O)-OH$

6. $CH_3-CH_2-C(=O)-H$ and $CH_3-C(=O)-CH_3$

Questions 7 to 9 refer to the isomers of oxygen-containing compounds.

    A. CH_3COOCH_3 B. $CH_3CH_2CH_2COOH$
    C. CH_3COCH_3 D. $CH_3CH_2CH_2CHO$

7. Which compound is an isomer of propanoic acid?

8. Which compound is an isomer of propanal?

9. Which compound is an isomer of butanone?

Nature's Chemistry

10. Which ester is an isomer of butanoic acid?

    A. ethyl ethanoate
    B. ethyl methanoate
    C. ethyl propanoate
    D. propyl ethanoate

11. Which compound is an isomer of hexanal?

    A. 2-methylbutanal
    B. 3-methylpentan-2-one
    C. 2,2-dimethylbutan-1-ol
    D. 3-ethylpentanal

In questions 12 to 17 decide whether each of the compounds

   A. does have an isomeric form
   B. does **NOT** have an isomeric form.

12. $CH_3CH_2CH_2OH$

13. C_2H_3Cl

14. chloroethane

15. C_3H_7Cl

16. CH_3CHCl_2

17. methanol

Nature's Chemistry

Test 2.7 Addition reactions (revision)

In questions 1 to 8 decide whether each of the hydrocarbons

- **A.** does undergo an addition reaction
- **B.** does **NOT** undergo an addition reaction.

1. octane
2. pentene
3. cyclopropane
4. cycloheptene
5. 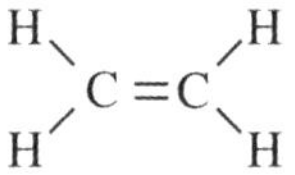 $CH_2=CH_2$ (ethene structure shown)
6. 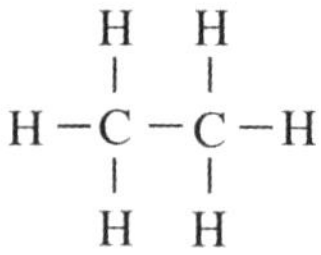 CH_3-CH_3 (ethane structure shown)
7. 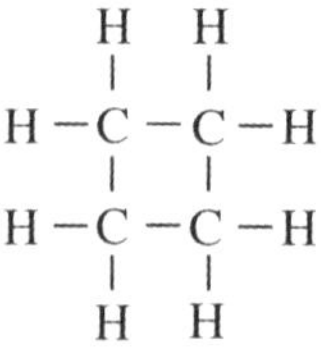 cyclobutane structure shown
8. 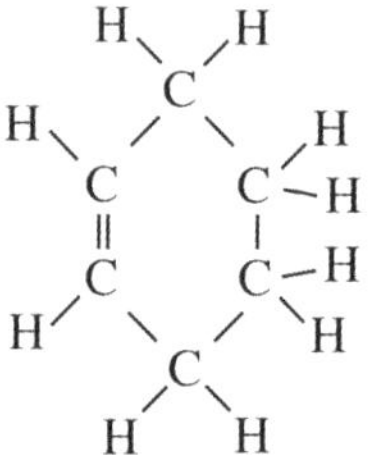 cyclohexene structure shown

9. A hydrocarbon, molecular formula C_5H_{10}, does **not** quickly decolourise bromine.

   Which hydrocarbon could it be?

   - **A.** pentane
   - **B.** cyclopentane
   - **C.** pentene
   - **D.** cyclopentene

10. When a molecule of the compound $CH_2=CH-CH=CH_2$ completely reacts with bromine, the number of molecules of bromine used would be

    - **A.** 1
    - **B.** 2
    - **C.** 3
    - **D.** 4.

11. What kind of reaction takes place when butene decolourises bromine?

    - **A.** hydrolysis
    - **B.** addition
    - **C.** oxidation
    - **D.** condensation

Nature's Chemistry

For questions 12 and 13, more than one answer should be given.

12. Which of the following represents the product of the reaction between ethene and bromine?

A.
$$\begin{array}{c} H \\ \diagdown \\ Br \end{array} C=C \begin{array}{c} H \\ \diagup \\ Br \end{array}$$

B.
$$\begin{array}{cc} H & Br \\ | & | \\ H-C-C-Br \\ | & | \\ H & H \end{array}$$

C.
$$\begin{array}{cc} Br & Br \\ | & | \\ Br-C-C-Br \\ | & | \\ Br & H \end{array}$$

D.
$$\begin{array}{cc} H & H \\ | & | \\ H-C-C-H \\ | & | \\ Br & H \end{array}$$

E.
$$\begin{array}{cc} Br & H \\ | & | \\ H-C-C-H \\ | & | \\ Br & H \end{array}$$

F.
$$\begin{array}{cc} H & Br \\ | & | \\ H-C-C-H \\ | & | \\ Br & H \end{array}$$

G.
$$\begin{array}{cc} Br & Br \\ | & | \\ Br-C-C-Br \\ | & | \\ Br & Br \end{array}$$

H.
$$\begin{array}{cc} H & H \\ | & | \\ Br-C-C-Br \\ | & | \\ H & H \end{array}$$

13. Which of the following represents the product of the reaction between propene and bromine?

A.
$$\begin{array}{ccc} H & H & H \\ | & | & | \\ H-C-C-C-H \\ | & | & | \\ Br & H & Br \end{array}$$

B.
$$\begin{array}{ccc} Br & H & H \\ | & | & | \\ Br-C-C-C-H \\ | & | & | \\ H & H & H \end{array}$$

C.
$$\begin{array}{c} Br \\ \diagdown \\ Br \end{array} C=C \begin{array}{cc} H & H \\ | & | \\ -C-H \\ | \\ H \end{array}$$

D.
$$\begin{array}{ccc} Br & H & H \\ | & | & | \\ H-C-C-C-H \\ | & | & | \\ H & Br & H \end{array}$$

E.
$$\begin{array}{c} H \\ \diagdown \\ Br \end{array} C=C \begin{array}{cc} Br & H \\ | & | \\ -C-H \\ | \\ H \end{array}$$

F.
$$\begin{array}{ccc} H & H & H \\ | & | & | \\ H-C-C-C-H \\ | & | & | \\ Br & Br & H \end{array}$$

G.
$$\begin{array}{ccc} H & H & H \\ | & | & | \\ Br-C-C-C-Br \\ | & | & | \\ H & H & H \end{array}$$

H.
$$\begin{array}{ccc} H & Br & H \\ | & | & | \\ H-C-C-C-Br \\ | & | & | \\ H & H & H \end{array}$$

Questions 14 and 15 refer to the reaction of butene with hydrogen?

14. What is the product of the reaction?

    A. ethane   B. propane   C. butane   D. hexane

15. What name is given to the reaction that takes place.

    A. hydration   B. hydrogenation   C. hydrolysis

16. Which hydrocarbon reacts with hydrogen to form hexane?

    A. propene   B. pentane   C. hexene   D. octane

Questions 17 and 18 refer to the reactions of ethyne.

Ethyne is an unsaturated hydrocarbon with the structure: H—C≡C—H
Ethyne can take part in addition reactions. Due to the carbon to carbon triple bond, the reactions can take place in two stages, each involving the breaking of a carbon to carbon covalent bond.

17. Ethyne can react with bromine.

    What is the full structural formula for the product of the second stage of the reaction?

    A.
    ```
    H   H
    |   |
    C = C
    |   |
    Br  Br
    ```

    B.
    ```
    Br  H
    |   |
    C = C
    |   |
    Br  H
    ```

    C.
    ```
       H   Br
       |   |
    H— C — C —H
       |   |
       Br  H
    ```

    D.
    ```
       Br  Br
       |   |
    H— C — C —H
       |   |
       Br  Br
    ```

Nature's Chemistry

18. Ethyne reacts with chlorine to give an unsaturated product **X**.
    Product **X** then reacts with hydrogen forming a saturated product **Y**.

    What is the full structural formula for product **Y**?

A.
$$\begin{array}{c} \text{Cl} \quad \text{H} \\ | \quad | \\ \text{H}-\text{C}-\text{C}-\text{H} \\ | \quad | \\ \text{Cl} \quad \text{H} \end{array}$$

B.
$$\begin{array}{c} \text{Cl} \quad \text{Cl} \\ | \quad | \\ \text{H}-\text{C}-\text{C}-\text{H} \\ | \quad | \\ \text{H} \quad \text{H} \end{array}$$

C.
$$\begin{array}{c} \text{Cl} \quad \text{Cl} \\ | \quad | \\ \text{H}-\text{C}-\text{C}-\text{H} \\ | \quad | \\ \text{Cl} \quad \text{Cl} \end{array}$$

D.
$$\begin{array}{c} \text{Cl} \quad \text{Cl} \\ | \quad | \\ \text{Cl}-\text{C}-\text{C}-\text{Cl} \\ | \quad | \\ \text{Cl} \quad \text{Cl} \end{array}$$

Test 2.8 Reactions involving alcohols

1. Ethanol can be made from ethene.

   What **two** names can be applied to the type of reaction that takes place?

   A. dehydration
   B. hydration
   C. condensation
   D. hydrolysis
   E. oxidation
   F. addition

Questions 2 to 8 refer to the reactions that take place when alcohols are passed over aluminium oxide as shown.

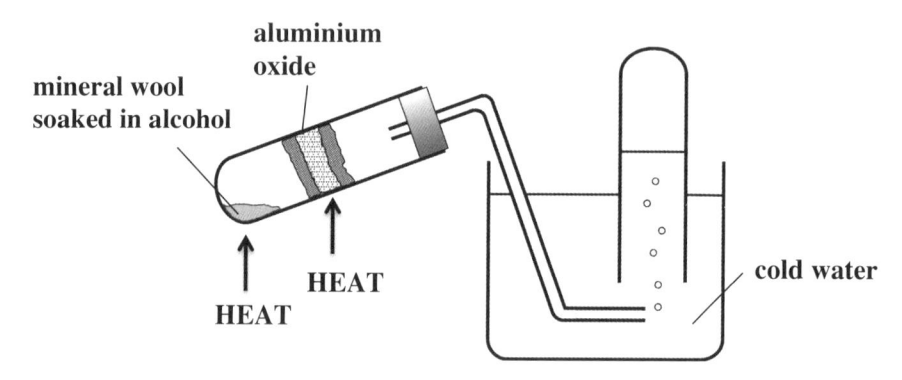

The elements of water are lost and a double carbon to carbon bond is formed. Depending on the position of the hydroxyl group, there can be one or two products.

2. The reaction taking place is an example of

   A. condensation
   B. dehydration
   C. dehydrogenation
   D. hydrolysis.

3. What is the product when ethanol is used in the reaction?

   A. ethane
   B. ethene
   C. ethanal
   D. ethanoic acid

4. What is the product when butan-1-ol is used in the reaction?

   A. but-1-ene
   B. but-2-ene
   C. a mixture of but-1-ene and but-2-ene

5. What is the product when butan-2-ol is used in the reaction?

    A. but-1-ene
    B. but-2-ene
    C. a mixture of but-1-ene and but-2-ene

6. What alcohol will produce pent-2-ene as the only product?

    A. pentan-1-ol
    B. pentan-2-ol
    C. pentan-3-ol

7. Which alcohol can produce, on dehydration, a pair of isomeric alkenes?

    A. propan-2-ol
    B. pentan-3-ol
    C. hexan-3-ol
    D. heptan-4-ol

8. What product(s) would be expected upon dehydration of the following alcohol ?

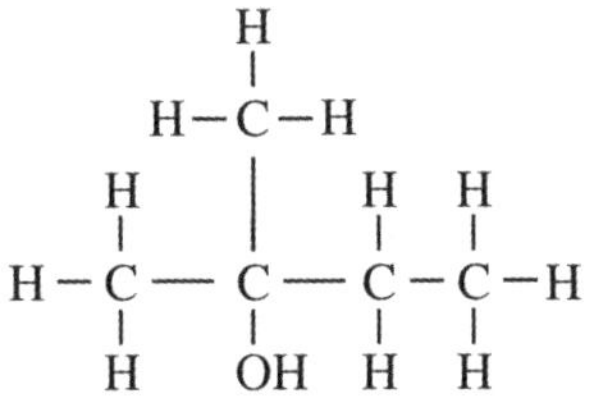

    $$\begin{array}{c} H \\ | \\ H-C-H \\ H \quad | \quad H \quad H \\ | \quad | \quad | \quad | \\ H-C-C-C-C-H \\ | \quad | \quad | \quad | \\ H \quad OH \quad H \quad H \end{array}$$

    A. 2-methylbut-2-ene only
    B. 2-methylbut-2-ene and 2-methylbut-1-ene
    C. 2-methylbut-1-ene only
    D. 3-methylbut-1-ene and 2-methylbut-1-ene

Test 2.9 — Structure of alcohols

The questions in this test refer to alcohols.

A.
$$CH_3-\underset{\underset{H}{|}}{\overset{\overset{OH}{|}}{C}}-CH_3$$

B.
$$CH_3-\underset{\underset{CH_3}{|}}{\overset{\overset{CH_3}{|}}{C}}-OH$$

C. Cyclohexane ring with H and OH on one carbon (cyclohexanol)

D. Cyclopentane ring with a CH$_3$ and OH on one ring carbon (1-methylcyclopentan-1-ol)

E. CH_3-OH

F.
$$CH_3-\underset{\underset{H}{|}}{\overset{\overset{CH_3}{|}}{C}}-CH_2-OH$$

G.
$$CH_3-\underset{\underset{CH_3}{|}}{\overset{\overset{CH_3}{|}}{C}}-CH_2-OH$$

H.
$$CH_3-\underset{\underset{CH_3}{|}}{\overset{\overset{CH_3}{|}}{C}}-\underset{\underset{CH_3}{|}}{\overset{\overset{H}{|}}{C}}-OH$$

I. $CH_3CH(OH)CH_2CH_3$

J. $CH_3CH_2C(CH_3)_2OH$

K. $CH_3CH(CH_3)CH(CH_3)OH$

L. ethanol

M. hexan-2-ol

N. 2-methylbutan-1-ol

O. 3-methylpentan-2-ol

P. 3-methylhexan-3-ol

1. Pick out **all** the alcohols that are **primary** alcohols.

2. Pick out **all** the alcohols that are **secondary** alcohols.

3. Pick out **all** the alcohols that are **tertiary** alcohols.

Test 2.10

Oxidation (i)

In questions 1 to 14 decide whether each of the alcohols can be oxidised to

A. an aldehyde **B.** a ketone **C.** neither.

1. CH_3-CH_2-OH

2. 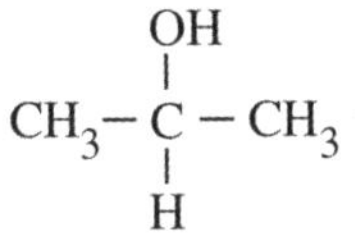 $CH_3-\underset{\underset{H}{|}}{\overset{\overset{OH}{|}}{C}}-CH_3$

3. $CH_3-\underset{\underset{CH_3}{|}}{CH}-CH_2-OH$

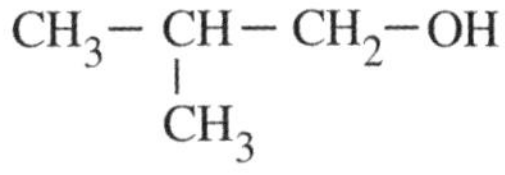

4. 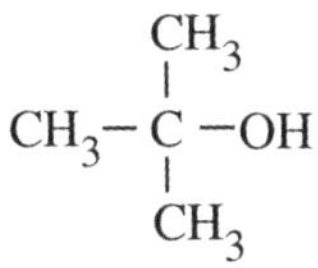 $CH_3-\underset{\underset{CH_3}{|}}{\overset{\overset{CH_3}{|}}{C}}-OH$

5. 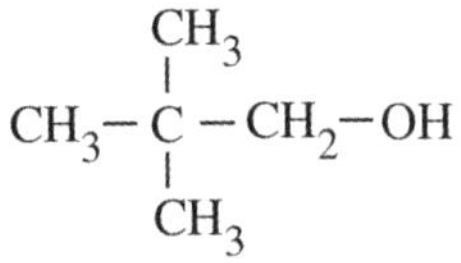 $CH_3-\underset{\underset{CH_3}{|}}{\overset{\overset{CH_3}{|}}{C}}-CH_2-OH$

6. 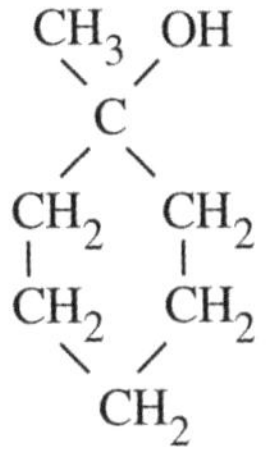 cycloheptane ring with CH_3 and OH on the same carbon

7. $CH_3CH_2CH_2OH$

8. $CH_3CH(OH)CH_2CH_3$

9. propan-1-ol

10. pentan-3-ol

11. 3-methylpentan-3-ol

12. 2-methylpentan-3-ol

13. 2-methylhexan-1-ol

14. cycloheptanol

Nature's Chemistry 61

In questions 15 to 22 decide whether each of the carbonyl compounds

    A.   can be easily oxidised to a carboxylic acid
    B.   **CANNOT** be easily oxidised to a carboxylic acid.

15. ethanol

16. propanone

17. pentanal

18. hexan-2-one

19.  $H-\underset{H}{\overset{}{C}}=O$

20. 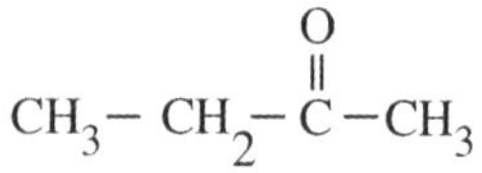 $CH_3-CH_2-\overset{\overset{O}{\|}}{C}-CH_3$

21. 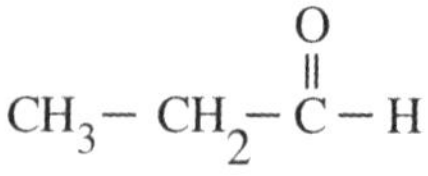 $CH_3-CH_2-\overset{\overset{O}{\|}}{C}-H$

22. 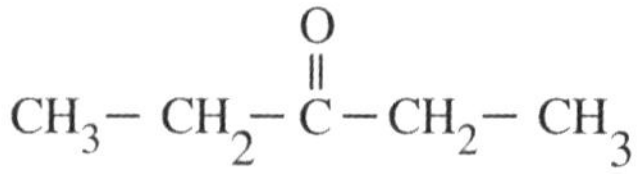 $CH_3-CH_2-\overset{\overset{O}{\|}}{C}-CH_2-CH_3$

Test 2.11

Oxidation (ii)

Questions 1 to 6 refer to the experiment shown.

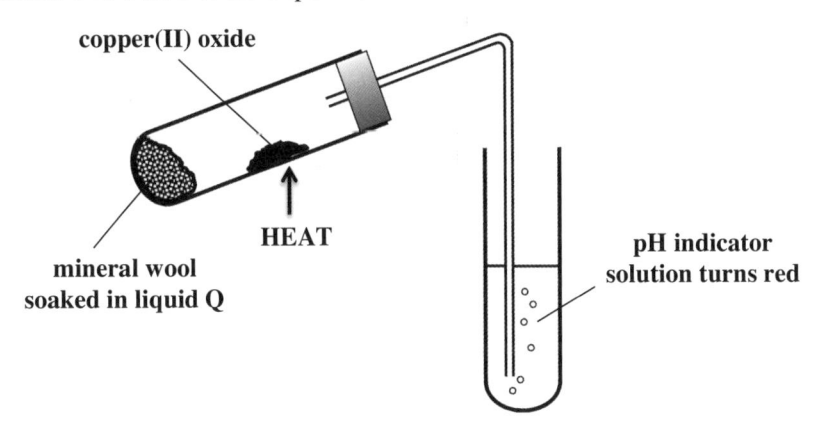

[Diagram: test tube containing mineral wool soaked in liquid Q and copper(II) oxide, being heated, with tube leading to pH indicator solution which turns red]

copper(II) oxide

mineral wool soaked in liquid Q

HEAT

pH indicator solution turns red

Decide whether each of the liquids

A. could be liquid **Q** B. could **NOT** be liquid **Q**.

1. propanone
2. paraffin
3. propanal
4. pentane
5. propan-1-ol
6. propan-2-ol

Questions 7 to 9 refer to the oxidation of carbonyl compounds.

A. propanone B. butanal
C. butanone D. propanal

7. What compound is formed by the oxidation of butan-2-ol?

8. What compound is oxidised to produce $CH_3CH_2CH_2COOH$?

9. What compound is formed by the oxidation of $CH_3CH_2CH_2OH$?

Nature's Chemistry 63

10. When the vapour of a liquid **X** is passed over heated copper(II) oxide, a reaction occurs and the vapour produced gives an orange precipitate with Benedict's solution.

    Which of the following could be **X**?

    A. propan-1-ol
    B. propan-2-ol
    C. propanal
    D. propanone

11. What type of reaction takes place when methanol is converted to methanal?

    A. condensation
    B. oxidation
    C. hydrolysis
    D. dehydration

12. Propan-1-ol is converted to propanal by warming it with potassium dichromate solution acidified with sulphuric acid.

    The purpose of the acid/dichromate mixture is to

    A. reduce the alcohol
    B. dehydrate the alcohol
    C. hydrate the alcohol
    D. oxidise the alcohol.

13. Which compound will produce a carboxylic acid on oxidation?

    A. CH_3-OH

    B. $CH_3-\underset{\underset{OH}{|}}{\overset{\overset{H}{|}}{C}}-CH_3$

    C. $CH_3-\underset{\underset{OH}{|}}{\overset{\overset{CH_3}{|}}{C}}-CH_3$

    D. $CH_3-\overset{\overset{O}{\|}}{C}-CH_3$

14. Bacterial oxidation of a solution of ethanol will result in the production of

    A. ethanoic acid
    B. ethene
    C. ethyl ethanoate
    D. ethane.

15. What compound is formed by the oxidation of propan-2-ol?

    A. CH_3CH_2CHO
    B. CH_3COCH_3
    C. CH_3CH_2COOH
    D. $CH_3CH_2CH_2OH$

16. Oxidation of 4-methylpentan-2-ol to the corresponding ketone results in the molecule

    A. losing 2 g per mole   B. gaining 2 g per mole
    C. gaining 16 g per mole   D. not changing in mass.

In questions 17 to 24 decide whether each of the reactions is an example of

    A. oxidation   B. reduction.

17. CH_3CH_2OH → CH_3COOH

18. $CH_3CH(OH)CH_3$ → CH_3COCH_3

19. $CH_3CH_2COCH_3$ → $CH_3CH_2CH(OH)CH_3$

20. CH_3CH_2CHO → CH_3CH_2COOH

21. C_2H_6O → C_2H_4O

22. C_7H_6O → $C_7H_6O_3$

23. $C_6H_{10}O$ → $C_6H_{12}O$

24. CH_3O_3 → CH_3O

Test 2.12

Simple esters – properties and reactions

In questions 1 to 6 decide whether each of the statements about methyl ethanoate is

A. TRUE **B.** FALSE.

1. It is soluble in water.
2. It has a characteristic smell.
3. It is a conductor of electricity.
4. It is flammable.
5. It is made up of molecules.
6. It turns Universal indicator solution red.

Questions 7 and 8 refer to the reaction of ethanoic acid and an alcohol in the presence of concentrated sulphuric acid.

7. The product is

   A. a hydrocarbon **B.** an ester
   C. a salt **D.** a carbohydrate.

8. The reaction can be considered to be an example of

   A. precipitation **B.** distillation
   C. condensation **D.** neutralisation.

9. The formation of ethanol from ethyl ethanoate is an example of

   A. condensation **B.** dehydration
   C. hydration **D.** hydrolysis.

10. What are the names of the alcohol and the acid produced on hydrolysis of the ester opposite?

    $$CH_3-\overset{O}{\underset{\|}{C}}-O-\underset{\underset{CH_3}{|}}{CH}-CH_3$$

    A. methanoic acid and propan-1-ol
    B. ethanoic acid and propan-1-ol
    C. methanoic acid and propan-2-ol
    D. ethanoic acid and propan-2-ol

Test 2.13　　　　　　　　　　Organic reactions

The questions in this test refer to types of reaction.

A.	addition	B.	hydration	C.	condensation
D.	oxidation	E.	hydrogenation	F.	hydrolysis
G.	dehydration	H.	dehydrogenation	I.	reduction

What type of reaction is each of the following?

For some of the questions in this test, **two** answers should be given.

1. ethene → ethane

2. methanol → methanal

3. butanal → butan-1-ol

4. butan-2-ol → butanone

5. but-1-ene → butan-1-ol + butan-2-ol

6. ethanol + butanoic acid → ethyl butanoate

7. propane → propene

8. ethanal → ethanoic acid

9. propan-2-ol → propene

10. methyl ethanoate → methanol + ethanoic acid

11. $CH_3-\underset{\underset{H}{|}}{\overset{\overset{OH}{|}}{C}}-CH_3$ → $CH_3-\overset{\overset{O}{\|}}{C}-CH_3$

12. $CH_3-CH=CH_2$ → $CH_3-\underset{\underset{H}{|}}{\overset{\overset{Cl}{|}}{C}}-\underset{\underset{H}{|}}{\overset{\overset{Cl}{|}}{C}}-H$

Nature's Chemistry

13. CH_3-CH_2-OH → $CH_3-\overset{\overset{O}{\|}}{C}-H$

14. $CH_3-CH_2-\overset{\overset{O}{\|}}{C}-OH$ → $CH_3-CH_2-\overset{\overset{O}{\|}}{C}-H$

15. CH_3-CH_2-OH → $CH_2=CH_2$

16. $H-\overset{\overset{O}{\|}}{C}-H$ → $H-\overset{\overset{O}{\|}}{C}-OH$

17. $H-\overset{\overset{O}{\|}}{C}-OH$ + CH_3-OH → $H-\overset{\overset{O}{\|}}{C}-O-CH_3$

18. CH_3-CH_3 → $CH_2=CH_2$

19. $CH_2=CH_2$ → CH_3-CH_2-OH

20. $H-\overset{\overset{O}{\|}}{C}-O-CH_2-CH_3$ → $H-\overset{\overset{O}{\|}}{C}-OH$ + CH_3-CH_2-OH

Nature's Chemistry

Test 2.14 Addition polymers (revision)

Questions 1 to 4 refer to the part of the polymer shown.

$$-\underset{H}{\overset{CH_3}{C}}-\underset{H}{\overset{H}{C}}-\underset{H}{\overset{CH_3}{C}}-\underset{H}{\overset{H}{C}}-\underset{H}{\overset{CH_3}{C}}-\underset{H}{\overset{H}{C}}-$$

1. How many repeating units are in the part of the polymer?

   A. 2   **B.** 3   **C.** 6   **D.** 9

2. What is the repeating unit?

   A. $-\underset{H}{\overset{H}{C}}-\underset{H}{\overset{H}{C}}-$

   B. $-\underset{H}{\overset{CH_3}{C}}-\underset{H}{\overset{H}{C}}-$

   C. $\underset{H}{\overset{CH_3}{C}}=\underset{H}{\overset{H}{C}}$

   D. $-\underset{H}{\overset{H}{C}}-\underset{H}{\overset{CH_3}{C}}-\underset{H}{\overset{H}{C}}-$

3. What is the name of the monomer?

   A. ethene   **B.** propane   **C.** propene   **D.** butane

4. What is the name of the polymer?

   A. polythene   **B.** poly(propene)
   C. poly(butene)   **D.** P.V.C.

5. Polyvinyl chloride is a polymer of vinyl chloride, $CH_2=CHCl$ (chloroethene).

   Which of the following is part of the formula for polyvinyl chloride?

   A. $-\underset{H}{\overset{Cl}{C}}-\underset{H}{\overset{Cl}{C}}-\underset{H}{\overset{Cl}{C}}-\underset{H}{\overset{Cl}{C}}-$

   B. $-\underset{Cl}{\overset{H}{C}}-\underset{H}{\overset{Cl}{C}}-\underset{Cl}{\overset{H}{C}}-\underset{H}{\overset{Cl}{C}}-$

   C. $-\underset{H}{\overset{H}{C}}-\underset{H}{\overset{Cl}{C}}-\underset{H}{\overset{H}{C}}-\underset{H}{\overset{Cl}{C}}-$

   D. $-\underset{H}{\overset{Cl}{C}}=\underset{}{\overset{Cl}{C}}-\underset{H}{\overset{Cl}{C}}=\underset{}{\overset{Cl}{C}}-$

6. Acrilan is an addition polymer of acrylonitrile. The structure of acrylonitrile is:

$$\text{H}_2\text{C}=\text{CH}-\text{CN}$$

Which of the following is part of the structure for Acrilan?

A.
```
    H   H   H   H
    |   |   |   |
  — C — C — C — C —
    |   |   |   |
    H   CN  H   CN
```

B.
```
    H   H           H   H
    |   |           |   |
  — C — C = N — C — C = N —
    |               |
    CH₃             CH₃
```

C.
```
    CN  H   CN  H
    |   |   |   |
  — C — C — C — C —
    |   |   |   |
    H   CN  H   CN
```

D.
```
    H   H           H   H
    |   |           |   |
  = C — C — N = C — C — N =
    |               |
    H               H
```

7. Which monomer could polymerise to give the polymer shown?

```
    CH₃ H   CH₃ H   CH₃ H
    |   |   |   |   |   |
  — C — C — C — C — C — C —
    |   |   |   |   |   |
    Cl  H   Cl  H   Cl  H
```

A.
```
    H   H   H
    |   |   |
    C = C — C — CH₃
    |       |
    H       Cl
```

B.
```
        H
        |
  CH₃ — C = C — CH₃
            |
            Cl
```

C.
```
            H   H
            |   |
  CH₃ — C = C
            |
            Cl
```

D.
```
            H
            |
  CH₃ — C = C
        |   |
        Cl  H
```

70   Nature's Chemistry

8. Which monomer could polymerise to give the polymer shown?

$$\begin{array}{cccccc} H & CH_3 & H & CH_3 & H & CH_3 \\ | & | & | & | & | & | \\ -C-&C-&C-&C-&C-&C- \\ | & | & | & | & | & | \\ H & CN & H & CN & H & CN \end{array}$$

A. CH_3 and CH_3 on one carbon, CN and H on the other (C=C)

B. H and CN on one carbon, CH_3 and CH_3 on the other (C=C)

C. H and CN on one carbon, CH_3 and H on the other (C=C)

D. H and H on one carbon, CH_3 and CN on the other (C=C)

9. Part of a polymer molecule is represented below.

$$\begin{array}{cccccccc} CH_3 & H & CH_3 & H & CH_3 & H & CH_3 & H \\ | & | & | & | & | & | & | & | \\ -C-&C-&C-&C-&C-&C-&C-&C- \\ | & | & | & | & | & | & | & | \\ H & CH_3 & H & CH_3 & H & CH_3 & H & CH_3 \end{array}$$

The monomer which gives rise to this polymer is

A. but-1-ene B. but-2-ene
C. methylpropene D. propene.

10. Part of a polymer molecule is shown.

$$\begin{array}{cccccccc} CH_3 & H & H & H & CH_3 & H & H & H \\ | & | & | & | & | & | & | & | \\ -C-&C-&C-&C-&C-&C-&C-&C- \\ | & | & | & | & | & | & | & | \\ H & H & H & H & H & H & H & H \end{array}$$

Which pair of alkenes was used as monomers?

A. ethene and propene B. ethene and but-1-ene
C. propene and but-1-ene D. ethene and but-2-ene

Test 2.15 — Condensation polymers

In questions 1 to 5 decide whether each of the polymers is made by

A. addition polymerisation   **B.** condensation polymerisation.

1. $-CH_2-CH(CH_3)-CH_2-CH(CH_3)-CH_2-CH(CH_3)-$

2. $-C(=O)-N(H)-(CH_2)_6-N(H)-C(=O)-(CH_2)_4-C(=O)-N(H)-(CH_2)_6-$

3. $-CF_2-CF_2-CF_2-CF_2-CF_2-CF_2-CF_2-CF_2-$

4. $-CH_2-CH(CN)-CH_2-CH(CN)-CH_2-CH(CN)-$

5. $-C(=O)-O-(CH_2)_2-O-C(=O)-(CH_2)_6-C(=O)-O-(CH_2)_2-O-$

Questions 6 and 7 refer to nylon. Part of the polymer chain is shown.

$$-\underset{\substack{\|\\O}}{C}-(CH_2)_4-\underset{\substack{\|\\O}}{C}-\underset{\substack{|\\H}}{N}-(CH_2)_4-\underset{\substack{|\\H}}{N}-\underset{\substack{\|\\O}}{C}-(CH_2)_4-\underset{\substack{\|\\O}}{C}-\underset{\substack{|\\H}}{N}-(CH_2)_4-\underset{\substack{|\\H}}{N}-$$

6. How many repeating units are in the part of the polymer shown?

   A. 1   B. 2   C. 3   D. 4

7. Which is (are) the monomers unit(s)?

   A. $HO-\underset{\substack{\|\\O}}{C}-(CH_2)_4-\underset{\substack{\|\\O}}{C}-\underset{\substack{|\\H}}{N}-(CH_2)_4-\underset{\substack{|\\H}}{N}-H$

   B. $HO-\underset{\substack{\|\\O}}{C}-(CH_2)_4-\underset{\substack{|\\H}}{N}-H$

   C. $HO-\underset{\substack{\|\\O}}{C}-(CH_2)_4-\underset{\substack{\|\\O}}{C}-OH$ $H-\underset{\substack{|\\H}}{N}-(CH_2)_4-\underset{\substack{|\\H}}{N}-H$

   D. $H-\underset{\substack{\|\\O}}{C}-(CH_2)_4-\underset{\substack{\|\\O}}{C}-H$ $HO-\underset{\substack{|\\H}}{N}-(CH_2)_4-\underset{\substack{|\\H}}{N}-OH$

8. A condensation polymer is made from the monomer shown.

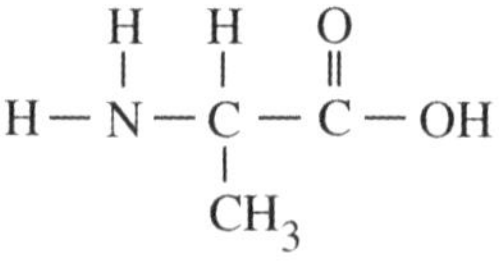

$$H-N(H)-C(H)(CH_3)-C(=O)-OH$$

Which of the following is part of the polymer chain?

A. 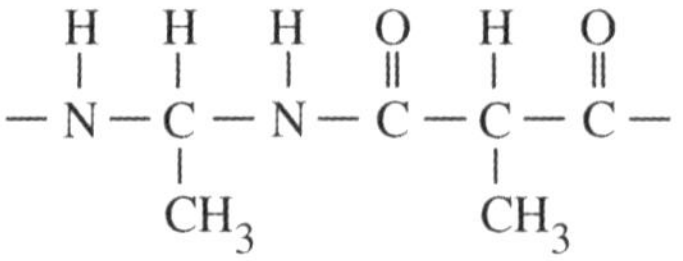 $-N(H)-C(H)(CH_3)-N(H)-C(=O)-C(H)-C(=O)-$

B. $-N(H)-C(H)(CH_3)-C(OH)(H)-N(H)-C(H)(CH_3)-C(H)(OH)-$

C. $-N(H)-C(H)(CH_3)-C(=O)-N(H)-C(H)(CH_3)-C(=O)-$

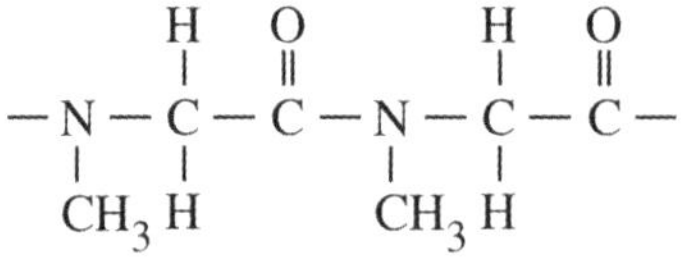

D. 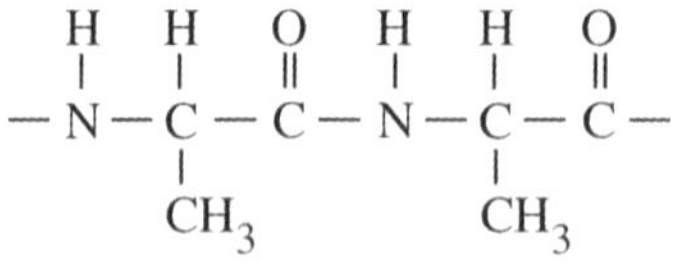 $-N(H)-C(H)(CH_3)-C(=O)-N(H)-C(H)(CH_3)-C(=O)-$

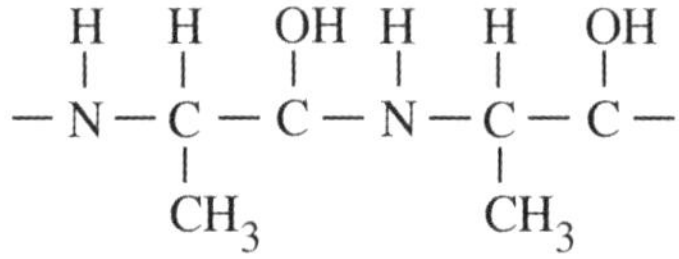

9. The following monomers can be used to prepare nylon–6,6.

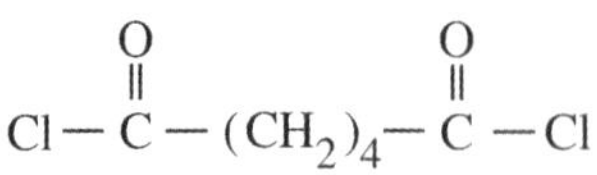

$$Cl-C(=O)-(CH_2)_4-C(=O)-Cl \qquad H_2N-(CH_2)_6-NH_2$$

Which molecule is released during the condensation reaction between these monomers?

A. HCl
B. H₂O
C. NH₃
D. HOCl

10. Which structure shows an amide link?

A. 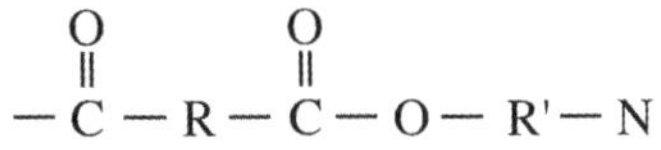
$$-\overset{O}{\underset{\|}{C}}-R-\overset{O}{\underset{\|}{C}}-O-R'-N$$

B. 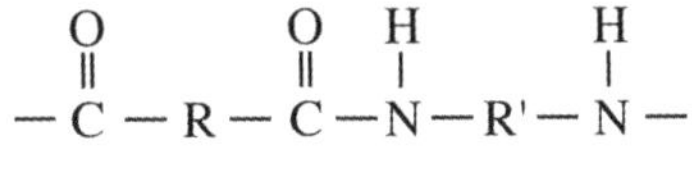
$$-\overset{O}{\underset{\|}{C}}-R-\overset{O}{\underset{\|}{C}}-\overset{H}{\underset{|}{N}}-R'-\overset{H}{\underset{|}{N}}-$$

C. $-O-R-O-R'-O-$

D. 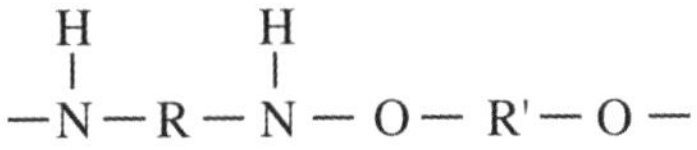 $-\overset{H}{\underset{|}{N}}-R-\overset{H}{\underset{|}{N}}-O-R'-O-$

11. Part of a polymer chain is shown.

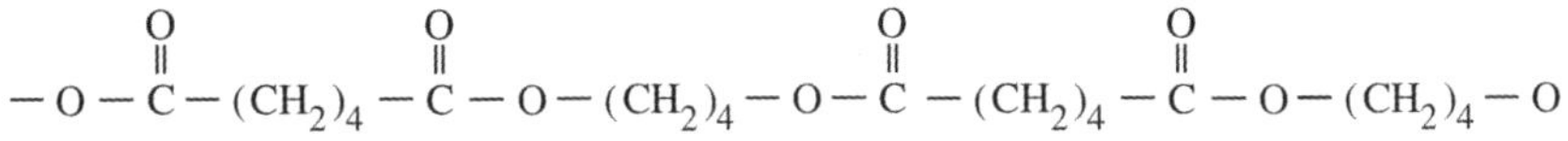

$$-O-\overset{O}{\underset{\|}{C}}-(CH_2)_4-\overset{O}{\underset{\|}{C}}-O-(CH_2)_4-O-\overset{O}{\underset{\|}{C}}-(CH_2)_4-\overset{O}{\underset{\|}{C}}-O-(CH_2)_4-O$$

Which compound, when added to the reactants during polymerisation, would stop the polymer chain from getting too long?

A. 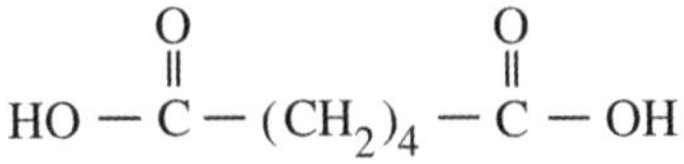 $HO-\overset{O}{\underset{\|}{C}}-(CH_2)_4-\overset{O}{\underset{\|}{C}}-OH$

B. 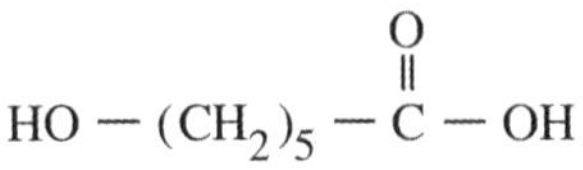 $HO-(CH_2)_5-\overset{O}{\underset{\|}{C}}-OH$

C. $HO-(CH_2)_6-OH$

D. CH_3-OH

Nature's Chemistry

Test 2.16 — Fats and oils

In questions 1 to 5 decide whether each of the statements is

A. TRUE   B. FALSE.

1. Fats and oils in the diet supply the body with energy.

2. Carbohydrates are a more concentrated source of energy than fats and oils.

3. Fats are likely to have relatively low melting points compared to oils.

4. Oils have a higher degree of unsaturation than fats.

5. Molecules in fats are packed more closely together than in oils.

6. Fats and oils can be classified as

    A. carbohydrates   B. acids
    C. esters   D. alcohols.

7. The breakdown of fats and oils produces glycerol and

    A. acids   B. alkanes
    C. alkenes   D. esters.

8. What is the structural formula for glycerol?

   A.
   CH_2OH
   |
   $CHOH$
   |
   CH_2OH

   B.
   CH_2OH
   |
   CH_2
   |
   CH_2OH

   C.
   CH_2OH
   |
   CH_2OH

   D.
   CH_2OH
   |
   $CHOH$
   |
   CH_2COOH

9. Glycerol can be obtained from a fat by

   A. oxidation
   B. condensation
   C. hydrolysis
   D. esterification.

10. The breakdown of fats and oils produces glycerol and fatty acids in the ratio of

    A. one mole to one mole
    B. one mole to two moles
    C. one mole to three moles
    D. one mole to four moles.

11. What type of reaction is represented by the following equation?

$$\begin{array}{l} CH_2-O-\overset{O}{\underset{\|}{C}}-C_{17}H_{35} \\ | \\ CH-O-\overset{O}{\underset{\|}{C}}-C_{17}H_{35} \\ | \\ CH_2-O-\overset{O}{\underset{\|}{C}}-C_{17}H_{35} \end{array} + 3H_2O \rightarrow \begin{array}{l} CH_2-OH \\ | \\ CH-OH \\ | \\ CH_2-OH \end{array} + 3C_{17}H_{35}COOH$$

   A. condensation
   B. hydrolysis
   C. oxidation
   D. dehydration

12. The conversion of linoleic acid, $C_{18}H_{32}O_2$, into stearic acid, $C_{18}H_{36}O_2$, is likely to be achieved by

    A. hydrogenation
    B. hydrolysis
    C. hydration
    D. dehydrogenation.

13. A vegetable oil is mixed with hydrogen under pressure at about 200 °C in the presence of a catalyst.

    The hydrogen

    A. dissolves in the oil without reacting
    B. combines with oxygen in the oil
    C. makes the unsaturated oil saturated
    D. makes the oil polymerise.

Nature's Chemistry

Questions 14 and 15 refer to kinds of reaction.

    A. hydrolysis   **B.** dehydration
    C. hydrogenation   **D.** condensation

14. What kind of reaction takes place during the breakdown of fats during digestion?

15. What kind of reaction takes place during the process by which some liquid oils can be converted into solid fats?

16. In the formation of 'hardened' fats from vegetable oil, the hydrogen

    A. causes cross-linking between chains
    B. causes hydrolysis to occur
    C. increases the carbon chain length
    D. reduces the number of carbon-carbon double bonds.

17. Fats have higher melting points than oils because comparing fats and oils

    A. fats have more hydrogen bonds
    B. fats have stronger van der Waals' forces between molecules
    C. fat molecules are more loosely packed
    D. fat molecules are more unsaturated.

18. Which of the following decolourises bromine solution least rapidly?

    A. palm oil
    B. hex-1-ene
    C. cod liver oil
    D. mutton fat

Test 2.17   Soaps, detergents and emulsifiers

In questions 1 to 4 decide whether each of the statements is

    A.   TRUE   B.   FALSE.

1. Ionic compounds are polar and generally soluble in water.

2. Hydrocarbons are polar and generally insoluble in oils.

3. Soaps are ionic compounds.

4. Part of the structure of a soap consists of a long chain of carbon atoms joined together by covalent bonds.

5. Which term can be used to classify a soap?

    A.   an amide   B.   an ester
    C.   a precipitate   D.   a salt

Questions 6 and 7 refer to the making of soap by treating a fat with sodium hydroxide solution.

$$\begin{array}{c} CH_2-O-\overset{\overset{O}{\|}}{C}-R \\ | \\ CH-O-\overset{\overset{O}{\|}}{C}-R^* \\ | \\ CH_2-O-\overset{\overset{O}{\|}}{C}-R^{**} \end{array} \rightarrow \begin{array}{c} CH_2-OH \\ | \\ CH-OH \\ | \\ CH_2-OH \end{array} + \begin{array}{c} R-\overset{\overset{O}{\|}}{C}-O^-Na^+ \\ \\ R^*-\overset{\overset{O}{\|}}{C}-O^-Na^+ \\ \\ R^{**}-\overset{\overset{O}{\|}}{C}-O^-Na^+ \end{array}$$

This reaction takes place in two stages.

6. What name is given to the type of reaction taking place at the first stage?

    A.   oxidation   B.   reduction
    C.   condensation   D.   hydrolysis

7. What name is given to the type of reaction taking place at the second stage?

    A.   neutralisation   B.   precipitation
    C.   condensation   D.   hydrolysis

Nature's Chemistry

Questions 8 and 9 refer to the representation of the structure of a soap/detergent:

```
       tail            head
```

A.	covalent	B.	ionic
C.	dissolves in oil	D.	dissolves in water
E.	hydrophilic	F.	hydrophobic

8. Identify the **three** terms that can be applied to the head of the soap structure.

9. Identify the **three** terms that can be applied to the tail of the soap structure.

Questions 10 to 13 refer to the shaking of soap with solutions.

Decide whether each of the solutions would produce

    A.   a lather   B.   a scum.

10. sodium chloride

11. calcium nitrate

12. magnesium chloride

13. potassium nitrate

14. The key difference between the structure of a soap and a detergent is that

    A.   the tail is shorter
    B.   the tail is longer
    C.   the negative ion has a different structure
    D.   the positive ion is different.

15. A soapless detergent forms a lather in hard water because

    A.   it forms a soluble calcium salt
    B.   it can neutralise the calcium ions present
    C.   it is hydrolysed by calcium salts
    D.   the precipitate is decomposed by the calcium ions present.

Questions 16 to 21 refer to emulsifiers.

Decide whether the molecule

A. could act as an emulsifier
B. could **NOT** act as an emulsifier.

16. 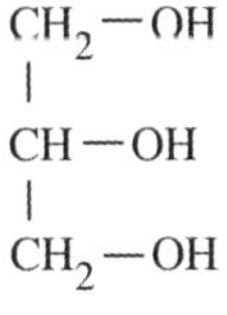
$$\begin{array}{c} CH_2-OH \\ | \\ CH-OH \\ | \\ CH_2-OH \end{array}$$

17. 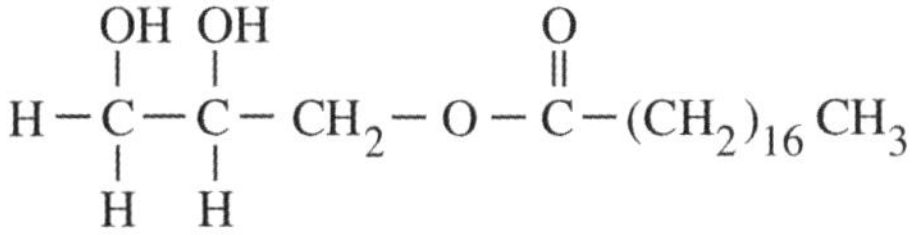
$$H-\underset{\underset{H}{|}}{\overset{\overset{OH}{|}}{C}}-\underset{\underset{H}{|}}{\overset{\overset{OH}{|}}{C}}-CH_2-O-\overset{\overset{O}{\|}}{C}-(CH_2)_{16}CH_3$$

18. 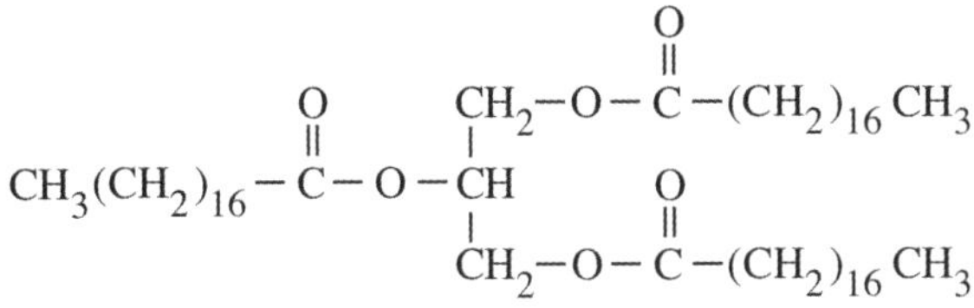
$$CH_3(CH_2)_{16}-\overset{\overset{O}{\|}}{C}-O-\underset{\underset{CH_2-O-\overset{\overset{O}{\|}}{C}-(CH_2)_{16}CH_3}{|}}{\overset{\overset{CH_2-O-\overset{\overset{O}{\|}}{C}-(CH_2)_{16}CH_3}{|}}{CH}}$$

19. 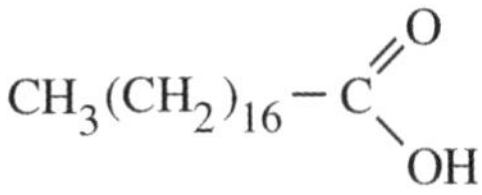
$$CH_3(CH_2)_{16}-C\underset{OH}{\overset{\nearrow O}{\diagdown}}$$

20. 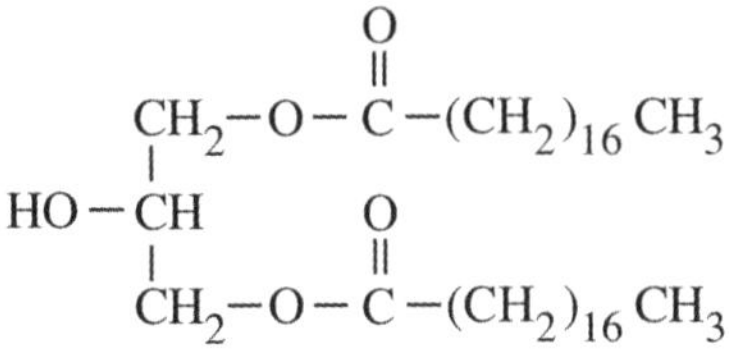
$$\begin{array}{c} CH_2-O-\overset{\overset{O}{\|}}{C}-(CH_2)_{16}CH_3 \\ | \\ HO-CH \\ | \\ CH_2-O-\overset{\overset{O}{\|}}{C}-(CH_2)_{16}CH_3 \end{array}$$

21. 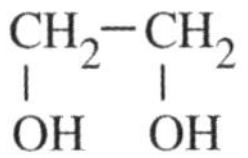
$$\begin{array}{cc} CH_2-CH_2 \\ | \quad\;\; | \\ OH \quad OH \end{array}$$

Nature's Chemistry

Test 2.18   Proteins

1. Proteins are substances which, in addition to carbon, hydrogen and oxygen, always contain

   A. phosphorus   B. nitrogen
   C. sulphur   D. calcium.

2.
$$H-\underset{\underset{CH_3}{|}}{\overset{\overset{CH_3}{|}}{C}}-\underset{\underset{H}{|}}{\overset{\overset{NH_2}{|}}{C}}-C\overset{\nearrow O}{\searrow OH}$$

   The molecule can be classified as

   A. an amino acid   B. an ester
   B a peptide   D. a protein.

Questions 3 and 4 refer to the breakdown of protein during digestion to give smaller molecules.

3. Which compound might be obtained by the breakdown of protein?

   A. glucose   B. glycerol
   C. stearic acid   D. amino-ethanoic acid

4. What name can be given to the type of reaction which takes place?

   A. dehydration   B. condensation
   C. hydrogenation   D. hydrolysis

5. Proteins can be denatured under acid conditions.

   During this process, the protein molecule

   A. changes shape   B. is dehydrated
   C. is neutralised   D. is polymerised.

6. In which kind of compound is nitrogen always present?

   A. enzymes   B. oils
   C. polyesters   D. carbohydrates

7. Which nitrogen containing compound could be a starting material for protein synthesis?

   A. $Pb(NO_3)_3$
   B. $(NH_4)_2SO_4$
   B $H_2N(CH_2)_4NH_2$
   D. $(CH_3)_2C(NH_2)COOH$

8. What type of chemical reaction takes place in the formation of proteins from amino acids?

   A. condensation
   B. hydration
   C. hydrolysis
   D. dehydration

9. Which of the following could represent part of a protein structure?

   A.
   ```
        H      H H
        |      | |
   — H — O — N — C — O —
               |
               H
   ```

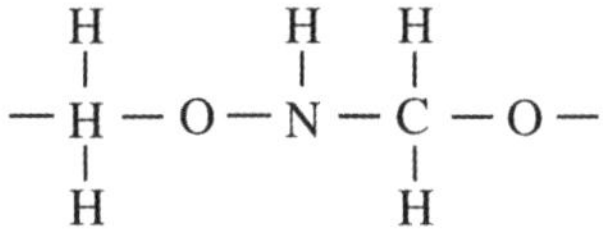

   B.
   ```
        O  H  H  O  H
        ||  |  |  ||  |
   — C — N — C — C — N —
                |
                H
   ```

   C.
   ```
       OH    H  OH
        |    |   |
   — C = N — C — C = N —
            |
            H
   ```

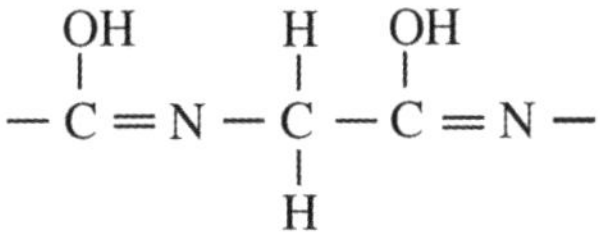

   D.
   ```
        H  O      H H
        |  ||     | |
   — C — C — O — N — C —
        |             |
        H             H
   ```

Nature's Chemistry   83

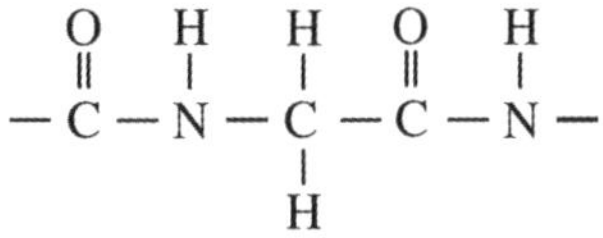

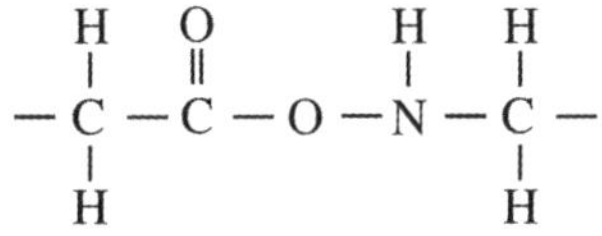

10. When two amino acids join together a peptide link is formed.

    Which of the following represents this process?

    A. 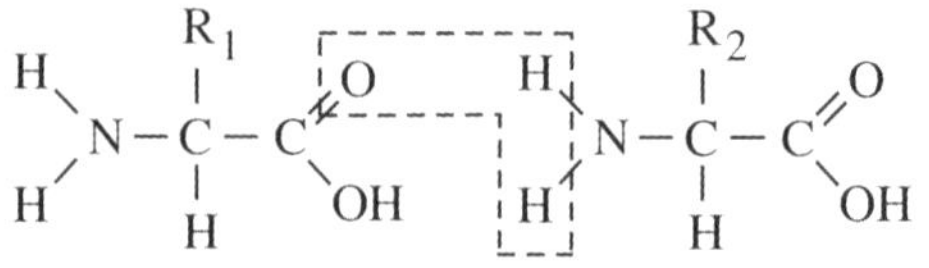

    B. 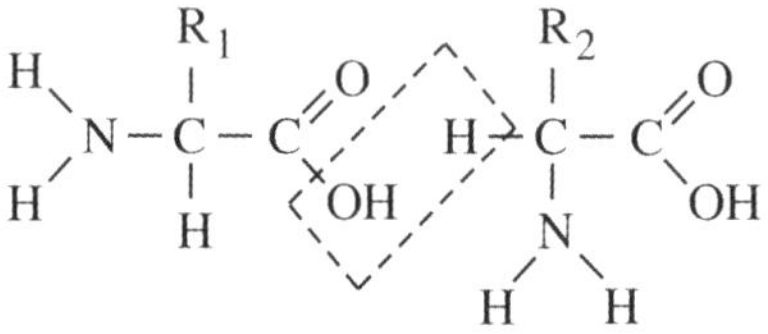

    C. 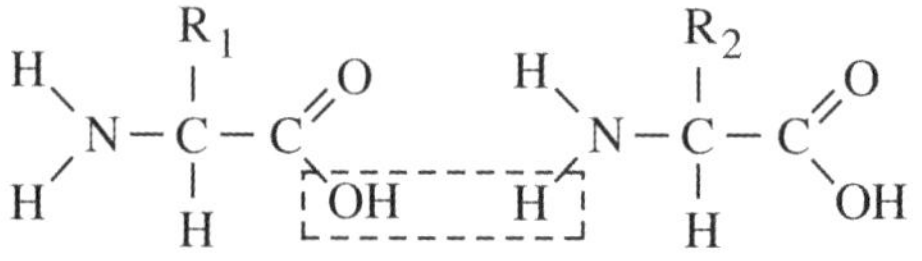

    D.

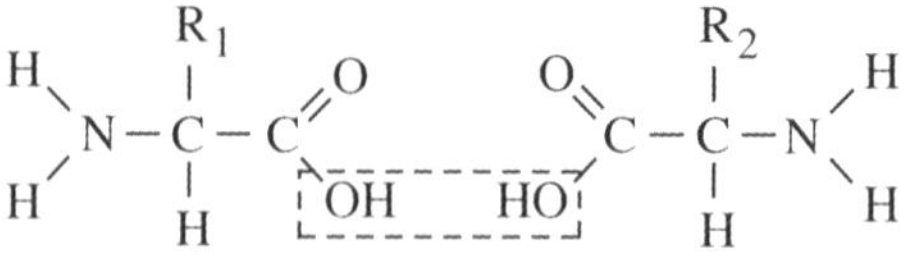

11. Which of the following is the most likely optimum temperature for human enzyme activity?

    A. close to 20 °C
    B. close to 40 °C
    C. close to 60 °C
    D. close to 80 °C

12. The monomer units used to construct enzyme molecules are

    A. esters
    B. amino acids
    C. fatty acids
    D. alkenes.

13. The graph shows how the rate of reaction varies with pH.

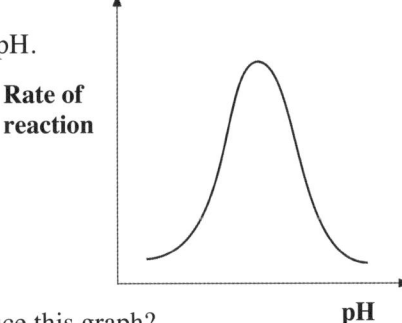

    Which reaction could produce this graph?

    A. the fermentation of glucose
    B. neutralisation of an acid by an alkali
    C. the combustion of sucrose
    D. the reaction of a metal with acid

14. The rate of hydrolysis of a protein, using an enzyme, was studied at different temperatures.

    Which graph could be obtained?

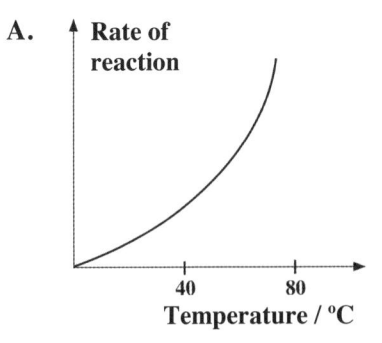

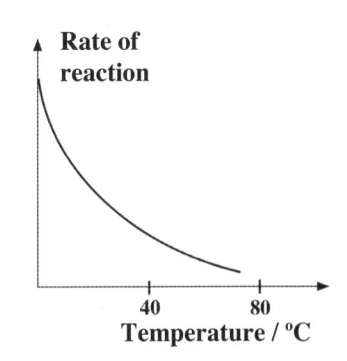

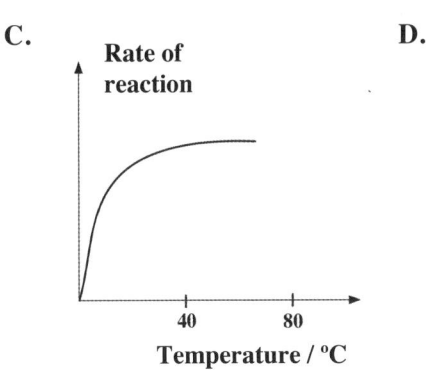

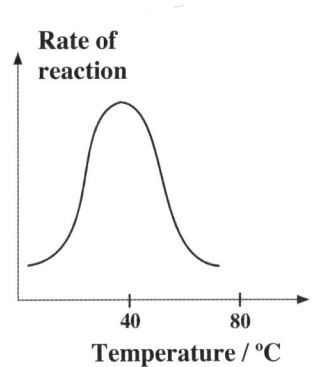

Test 2.19 Functional groups (ii)

Some carbon compounds have more than one functional group.

A.	alcohol	**B.**	aldehyde	**C.**	amide
D.	amino (or amine)	**E.**	ketone	**F.**	ester
G.	carboxylic acid				

What are the functional groups in each of the following compounds?

For the questions in this test, **two** answers should be given.

1.

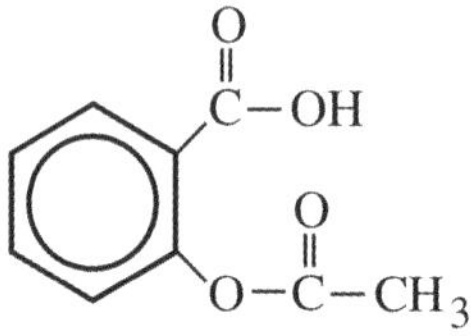

2.

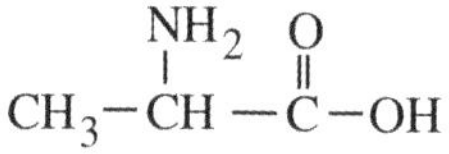

3.

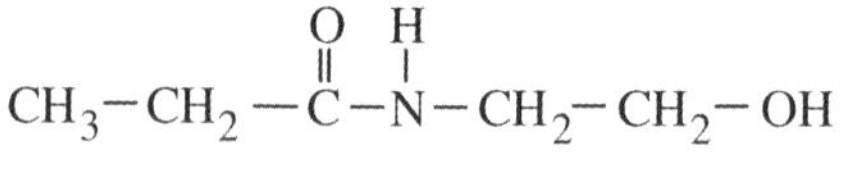

4.

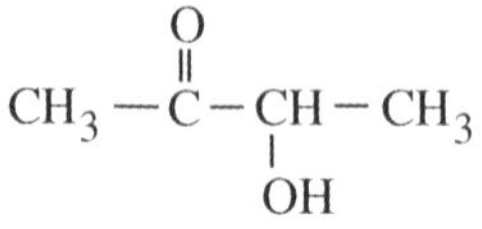

5.

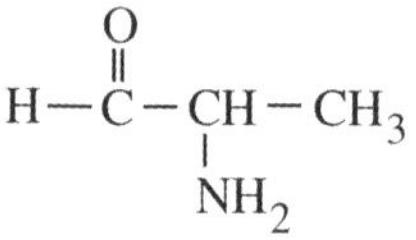

6.

w

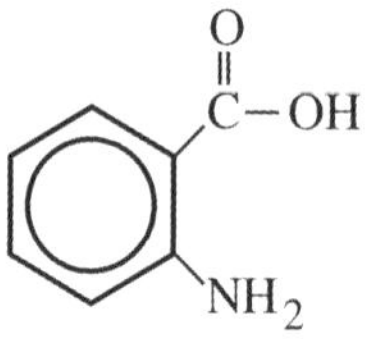

7.

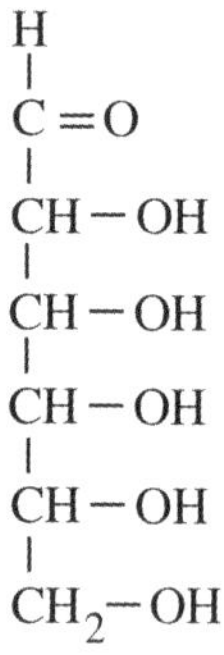

8.

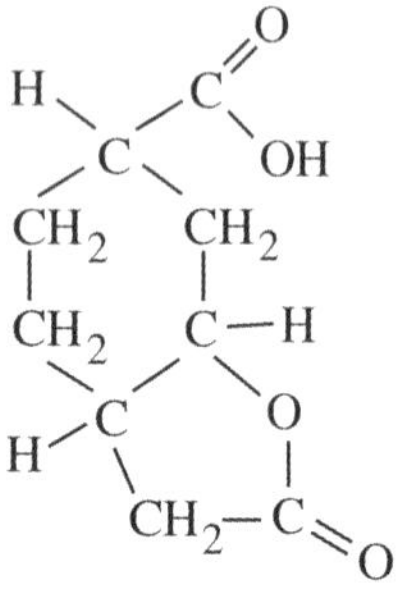

Test 2.20 Everyday chemistry

In questions 1 to 4 decide whether each of the statements is

A. TRUE B. FALSE.

1. Molecules with hydroxyl groups (-OH) are likely to be soluble in water.

2. Molecules made up of carbon and hydrogen atoms only are likely to be soluble in fats and oils.

3. Molecules with hydroxyl groups (-OH) are likely to be soluble in hexane.

4. Molecules made up of carbon and hydrogen atoms only are likely to be soluble in dilute ethanoic acid.

Questions 5 to 10 refer to the likely properties of carbon compounds.

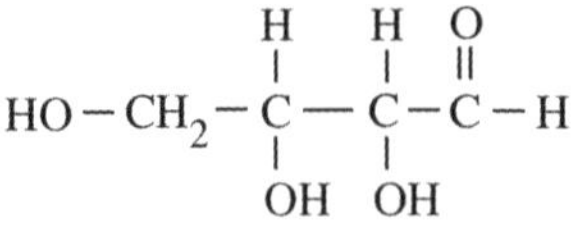

erythrose

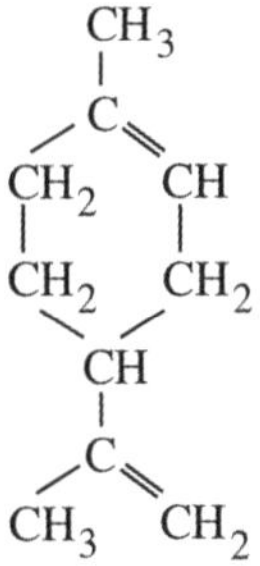

limonene

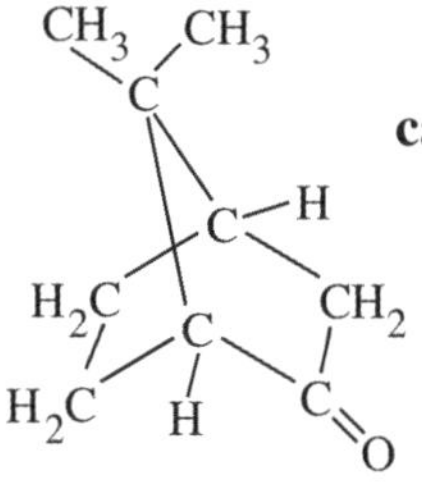

camphor

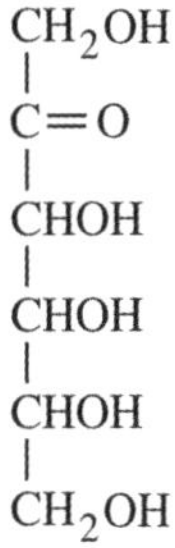

fructose

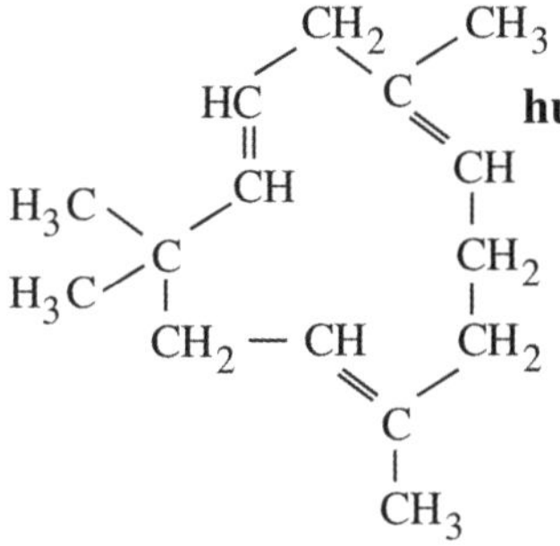

humulene

Decide whether each of the statements is

    A.   TRUE   **B.**   FALSE.

5. Erythrose is more likely to be soluble in ethanol than in hexane.

6. Limonene is more likely to be soluble in water than in octane.

7. Camphor is more likely to be soluble in propanone than in water.

8. Fructose is more likely to be soluble in propanal than in water.

9. Humulene is more likely to be soluble in ethanol than in cyclohexene.

10. Limonene is likely to be more volatile than humulene.

Questions 11 to 13 refer to two properties of carbon compounds.

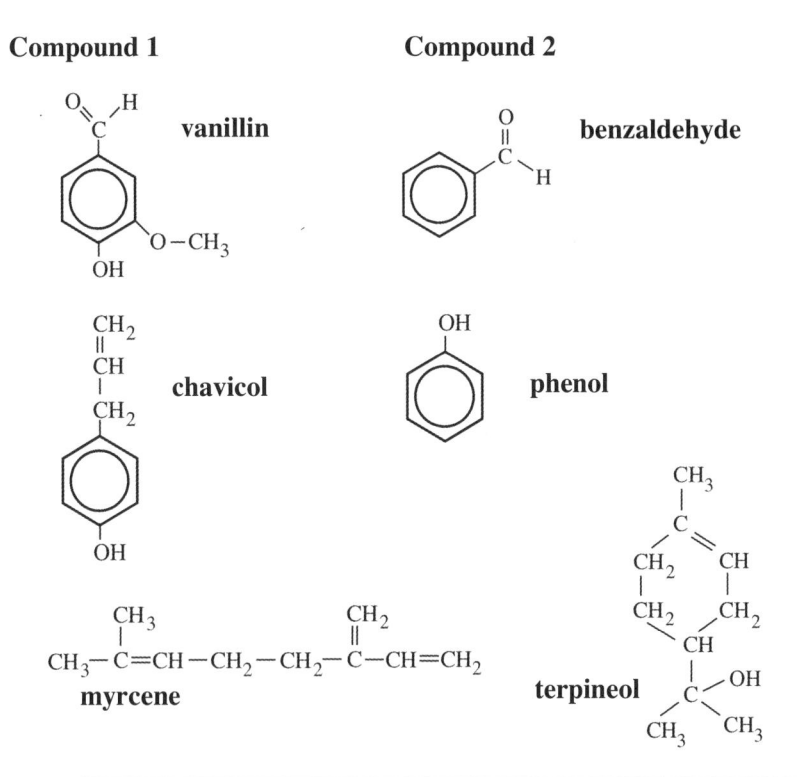

Compound 1

vanillin

chavicol

myrcene

Compound 2

benzaldehyde

phenol

terpineol

	Compound 1	**Compound 2**
A.	likely to be more soluble in water	likely to be more volatile
B.	likely to be more soluble in water	likely to be less volatile
C.	likely to be less soluble in water	likely to be more volatile
D.	likely to be less soluble in water	likely to be less volatile

11. Which line in the table compares the likely properties when compound 1 is vanillin and compound 2 is benzaldehyde?

12. Which line in the table compares the properties when compound 1 is chavicol and compound 2 is phenol?

13. Which line in the table compares the properties when compound 1 is myrcene and compound 2 is terpineol?

Nature's Chemistry

Test 2.21   Terpenes

In questions 1 to 6 decide whether each of the terms

    A.   can be used to describe the structure of isoprene.
    B.   **CANNOT** be used to describe the structure of isoprene.

1. aldehyde
2. ester
3. ketone
4. hydrocarbon
5. saturated
6. unsaturated

7. What is the molecular formula for isoprene?

    A. C_4H_6 **B.** C_4H_8 **C.** C_5H_8 **D.** C_5H_{10}

Questions 8 to 10 refer to the number of isoprene units used to form terpenes.

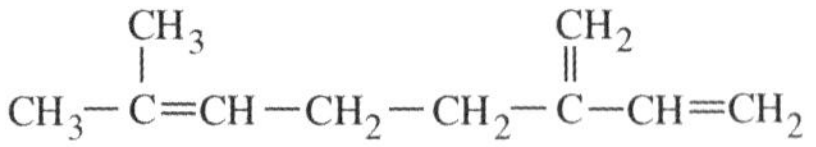

myrcene

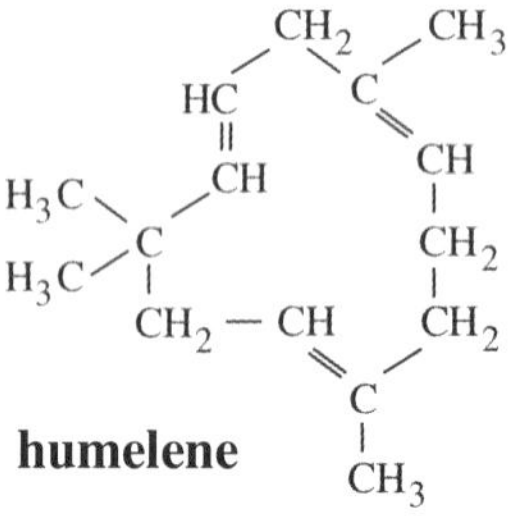

humelene

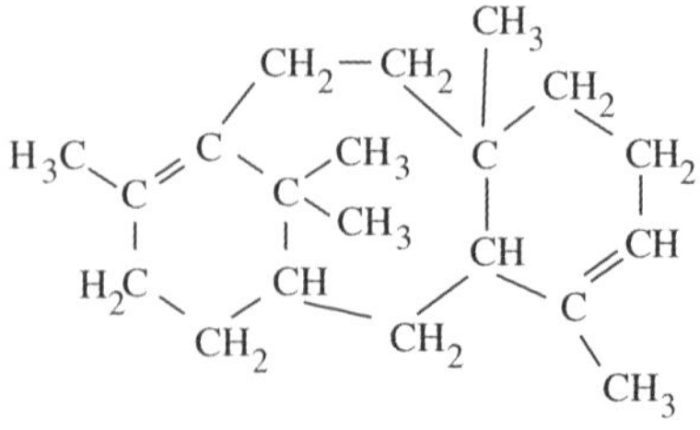

taxadiene

8. How many isoprene units were used to form a myrcene molecule?

    A. 2   **B.** 3   **C.** 4   **D.** 5

9. How many isoprene units were used to form a humelene molecule?

    A. 2   **B.** 3   **C.** 4   **D.** 5

10. How many isoprene units were used to form a taxadiene molecule?

    A. 2   **B.** 3   **C.** 4   **D.** 5

Questions 11 to 15 refer to the a functional group in oxidised terpenes known as terpenoids.

A. alcohol
B. aldehyde
C. carboxylic acid
D. ester
E. ketone

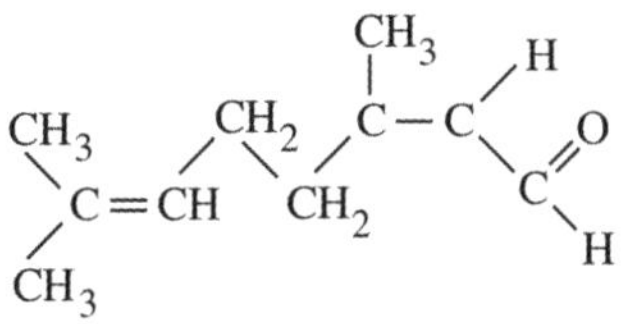

molecule 1

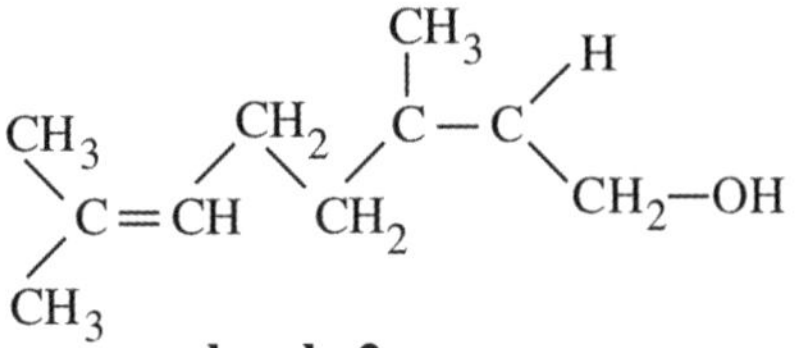

molecule 2

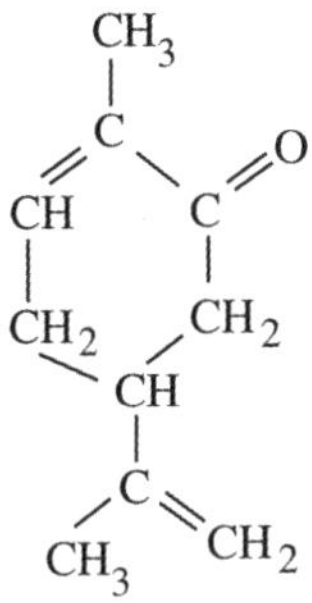

molecule 3

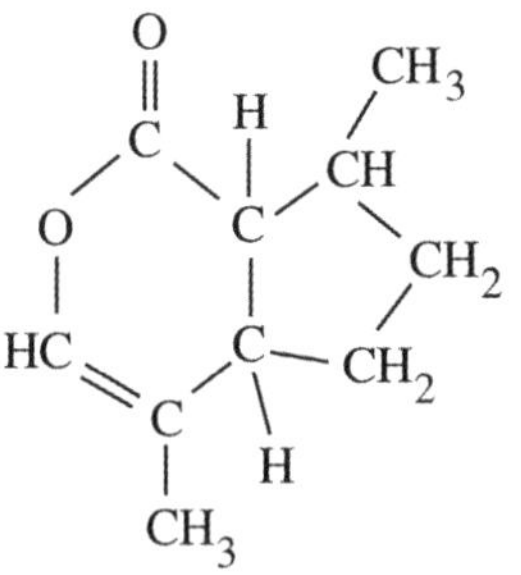

molecule 4

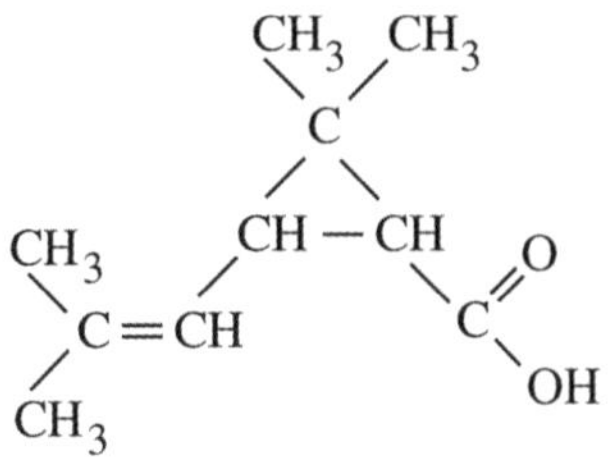

molecule 5

11. What is the name of the functional group in molecule 1?

12. What is the name of the functional group in molecule 2?

13. What is the name of the functional group in molecule 3?

14. What is the name of the functional group in molecule 4?

15. What is the name of the functional group in molecule 5?

Nature's Chemistry

Test 2.22 — Free radicals

In questions 1 to 8 decide whether each of the atoms or groups of atoms

    A.   represents a free radical
    B.   does **NOT** represent a free radical.

1. a hydrogen atom, H

2. an oxygen molecule, O_2

3. a chlorine atom, Cl

4. a methyl group, CH_3

5. an ethane molecule, C_2H_6

6. an amine group, NH_2

7. a hydroxyl group, OH

8. a hydrogen bromide molecule, HBr

In questions 9 to 11 decide whether each of the statements is

    A.   TRUE   **B.**   FALSE.

9. In a free radical, the atom (or atoms) always has a stable arrangement of electrons.

10. Ultra-violet radiation can result in skin damage due to the formation of free radicals.

11. Sun-block products effectively reduce skin damage mainly due to their ability to provide free radical scavengers.

Questions 12 to 17 refer to the light-initiated free radical chain reaction between hexane and bromine.

Decide whether each equation represents

    A. an initiation step
    B. a propagation step
    C. a termination step.

12. $Br^\bullet + Br^\bullet \rightarrow Br_2$

13. $C_6H_{13}^\bullet + Br_2 \rightarrow C_6H_{13}Br + Br^\bullet$

14. $C_6H_{13}^\bullet + C_6H_{13}^\bullet \rightarrow C_{12}H_{26}$

15. $Br_2 \rightarrow Br^\bullet + Br^\bullet$

16. $Br^\bullet + C_6H_{14} \rightarrow C_6H_{13}^\bullet + HBr$

17. $Br^\bullet + C_6H_{13}^\bullet \rightarrow C_6H_{13}Br$

Questions 18 to 20 refer to the light-initiated free radical chain reaction between hydrogen and chlorine.

For two of the questions, more than one answer should be given.

    A. $Cl_2 \rightarrow Cl^\bullet + Cl^\bullet$

    B. $H^\bullet + Cl_2 \rightarrow HCl + Cl^\bullet$

    C. $H^\bullet + Cl^\bullet \rightarrow HCl$

    D. $Cl^\bullet + Cl^\bullet \rightarrow Cl_2$

    E. $Cl^\bullet + H_2 \rightarrow HCl + H^\bullet$

    F. $H^\bullet + H^\bullet \rightarrow H_2$

18. Identify the initiation step(s).

19. Identify the propagation step(s).

20. Identify the termination step(s).

Test 3.1

The mole (revision)

1. What is the mass, in grams, of 0.25 mol of carbon dioxide?

    A. 7 B. 11 C. 15 D. 19

2. How many moles are contained in 1 g helium?

    A. 0.1 B. 0.25 C. 0.4 D. 1.0

3. How many moles of potassium carbonate must be dissolved to make 200 cm³ of 2 mol l⁻¹ solution?

    A. 0.02 B. 0.04 C. 0.2 D. 0.4

4. What is the concentration, in mol l⁻¹, of a solution which contains 0.5 mol of hydrogen chloride dissolved in 200 cm³ of solution?

    A. 0.2 B. 0.5 C. 1.0 D. 2.5

5. What volume of a 0.2 mol l⁻¹ lithium chloride solution contains 1 mol?

    A. 200 cm³ B. 500 cm³ C. 1 litre D. 5 litres

6. What mass of sodium carbonate, in grams, is required to make 50 cm³ of 0.1 mol l⁻¹ solution?

    A. 0.53 B. 1.06 C. 5.3 D. 10.6

7. What is the concentration of a solution, in mol l⁻¹, containing 4 g of sodium hydroxide in 100 cm³ of water?

    A. 0.01 B. 0.4 C. 1 D. 4

8. 0.5 mol of copper(II) chloride and 0.5 mol of copper(II) sulphate are dissolved in water and made up to 500 cm³ of solution.

    What is the concentration, in mol l⁻¹, of Cu^{2+} (aq) ions in the solution?

    A. 0.5 B. 1.0 C. 2.0 D. 4.0

Test 3.2

More on the mole

1. Which solid contains the greatest number of atoms?

   A. 20 g of carbon
   B. 20 g of calcium
   C. 20 g of magnesium
   D. 20 g of sulphur

2. Which gas contains the smallest number of molecules?

   A. 100 g of fluorine
   B. 100 g of nitrogen
   C. 100 g of oxygen
   D. 100 g of hydrogen

3. Which gas contains the greatest number of molecules?

   A. 0.10 g of hydrogen gas
   B. 0.17 g of ammonia gas
   C. 0.32 g of methane gas
   D. 0.16 g of oxygen gas

4. Which gas contains the greatest number of atoms?

   A. 1 g of hydrogen
   B. 32 g of oxygen
   C. 20.2 g of neon
   D. 39.9 g of argon

5. 160.4 g of calcium contains as many atoms as

   A. 28 g of carbon
   B. 92 g of sodium
   C. 54 g of aluminium
   D. 310 g of phosphorus.

6. 97.2 g of magnesium contains twice as many atoms as

   A. 9 g of beryllium
   B. 2 mol of oxygen molecules
   C. 2 mol of calcium
   D. 48 g of carbon.

7. Which of the following contains the greatest number of atoms?

   A. 12 g of carbon
   B. 9 g of hydrogen oxide
   C. 16 g of oxygen
   D. 14 g of carbon monoxide

8. Which of the following contains the smallest number of hydrogen atoms?

   A. 17 g of ammonia (NH_3)
   B. 16 g of methane (CH_4)
   C. 36 g of water (H_2O)
   D. 14 g of ethene (C_2H_4)

Chemistry in Society

9. 4 g of sodium hydroxide contains the same number of ions as

   A. 17 g of sodium nitrate
   B. 10.01 g of calcium carbonate
   C. 16.41 g of calcium nitrate
   D. 10.6 g of sodium carbonate.

10. Which of the following contains the fewest ions?

    A. 100 g of magnesium nitrate
    B. 100 g of ammonium bromide
    C. 100 g of calcium hydroxide
    D. 100 g of lithium carbonate

11. What is the amount, in moles, of oxygen atoms in 0.5 mol of carbon dioxide?

    A. 0.25   B. 0.5   C. 1   D. 2

12. How many moles of ions are in one mole of copper(II) phosphate?

    A. 2   B. 3   C. 4   D. 5

13. 36 g of hydrogen oxide contains

    A. 2 mol of hydrogen atoms
    B. 4 mol of hydrogen atoms
    C. 1 mol of atoms
    D. 2 mol of atoms.

14. Which of the following is **not** found in 1 g of hydrogen gas?

    A. 1 mol of electrons
    B. 1 mol of atoms
    C. 1 mol of protons
    D. 1 mol of molecules

15. In which of the following pairs do the gases contain the same number of atoms of oxygen?

    A. 1 mol of oxygen and 1 mol of carbon monoxide
    B. 1 mol of oxygen and 0.5 mol of carbon dioxide
    C. 0.5 mol of oxygen and 1 mol of carbon dioxide
    D. 1 mol of oxygen and 1 mol of carbon dioxide

16. Which of the following contains one mole of neutrons?

    A. 1 g of $^{12}_{6}C$
    B. 1 g of $^{16}_{8}O$
    C. 2 g of $^{3}_{1}H$
    D. 2 g of $^{4}_{2}He$

17. A mixture of magnesium chloride and magnesium sulphate is known to contain 0.6 mol of chloride ion and 0.2 mol of sulphate ion.

    How many moles of magnesium ions are present?

    A. 0.4   B. 0.5   C. 0.8   D. 1.0

18. A mixture of sodium chloride and sodium sulphate is known to contain 0.5 mol of sodium ion and 0.2 mol of chloride ion.

    How many moles of sulphate ions are present?

    A. 0.15   B. 0.20   C. 0.25   D. 0.30

19. A mixture of sodium chloride and sodium sulphate is known to contain 0.6 mol of chloride ion and 0.2 mol of sulphate ion.

    How many moles of sodium ions are present?

    A. 0.4   B. 0.5   C. 0.8   D. 1.0

20. A mixture of calcium carbonate and magnesium carbonate is known to contain 0.4 mol of calcium ion and 0.5 mol of carbonate ion.

    How many moles of magnesium ions are present?

    A. 0.1   B. 0.4   C. 0.5   D. 0.9

21. A mixture of magnesium bromide and magnesium sulphate is known to contain 3 mol of magnesium ion and 4 mol of bromide ion.

    How many moles of sulphate ions are present?

    A. 1   B. 2   C. 3   D. 4

22. A mixture of sodium sulphate and copper(II) sulphate is known to contain 3 mol of sulphate ion and 1 mol of copper ion.

    How many moles of sodium ions are present?

    A. 1   B. 2   C. 3   D. 4

Test 3.3   The Avogadro Constant (i)

For the questions in this test decide whether each of the statements is

    A.   TRUE   B.   FALSE.

Questions 1 to 17 refer to the Avogadro Constant.

The Avogadro Constant is the same as the number of:

1. atoms in 127 g of copper
2. atoms in 31 g of phosphorus
3. atoms in 0.5 mol of chlorine
4. atoms in 16 g of oxygen
5. molecules in 56 g of carbon monoxide
6. molecules in 0.25 mol of sulphur trioxide
7. molecules in 6 g of water
8. molecules in 2 g of hydrogen
9. ions in 1 litre of sodium chloride solution, concentration 1 mol l^{-1}
10. ions in 0.5 mol of sodium oxide
11. ions in 107 g of ammonium chloride
12. ions in 20 g of sodium hydroxide
13. lithium ions in 500 cm^3 of 2 mol l^{-1} lithium chloride solution
14. sulphate ions in 1 litre of 1 mol l^{-1} sulphuric acid
15. electrons in 0.5 mol of helium atoms
16. protons in 2 g of sulphur
17. neutrons in 2 g of nitrogen

Questions 18 to 20 refer to 17 g of ammonia, 2 g of hydrogen and 71 g of chlorine.

18. Each occupies the same volume.

19. Each contains approximately 6×10^{23} atoms.

20. Each contains approximately 6×10^{23} molecules.

Questions 21 to 26 refer to carbon dioxide gas.

21. Approximately, the mass of 6×10^{23} molecules is 44 g.

22. One molecule is 44 times as heavy as a molecule of hydrogen.

23. 44 g occupy the same volume as 32 g oxygen.

24. 44 g of the gas contains the same number of atoms as 20 g of neon.

25. 44 g of the gas contains approximately 6×10^{23} carbon atoms.

26. 44 g of the gas contains the same number of molecules as 1 g of hydrogen.

Chemistry in Society

Test 3.4   The Avogadro Constant (ii)

See note to teachers / students at front of the book.

For the questions in this test take the Avogadro Constant to be 6×10^{23} mol^{-1}.

1. How many atoms are in 46 g of sodium?

   A. 1×10^{23} **B.** 3×10^{23}
   C. 6×10^{23} **D.** 1.2×10^{24}

2. How many molecules are in 3.2 g of methane?

   A. 1×10^{23} **B.** 1.2×10^{23}
   C. 6×10^{23} **D.** 3×10^{24}

3. How many atoms are in 0.5 mol of fluorine?

   A. 1.5×10^{23} **B.** 3×10^{23}
   C. 6×10^{23} **D.** 1.2×10^{24}

4. How many atoms are in 0.44 g of carbon dioxide?

   A. 1.8×10^{22} **B.** 6×10^{22}
   C. 1.2×10^{23} **D.** 1.8×10^{23}

5. How many ions in 9.9 g of copper(I) chloride?

   A. 6×10^{22} **B.** 1.2×10^{23}
   C. 1.8×10^{23} **D.** 2.4×10^{23}

6. The molecular formula for a gas is X_3.
   How many **X** atoms will be present in 0.25 mol of X_3?

   A. $0.5 \times 6 \times 10^{23}$ **B.** $0.75 \times 6 \times 10^{23}$
   C. $1 \times 6 \times 10^{23}$ **D.** $3 \times 6 \times 10^{23}$

7. How many protons are in 120 g of carbon?

   A.   60   **B.** 1×10^{24}
   C. 6×10^{24} **D.** 3.6×10^{25}

8. Deuterium ($^{2}_{1}$H) is a heavy isotope of hydrogen.

   How many neutrons are in 10 g of deuterium atoms?

   A. 3×10^{23}
   C. 3×10^{24}
   B. 6×10^{23}
   D. 6×10^{24}

9. What is the mass, in grams, of one sodium atom?

   A. 6×10^{23}
   C. 3.8×10^{-23}
   B. 6×10^{-23}
   D. 3.8×10^{-24}

10. What is the mass, in grams, of 100 molecules of hydrogen?

    A. 6×10^{23}
    C. 3.33×10^{-22}
    B. 1.66×10^{-22}
    D. 1.2×10^{-23}

11. Fullerene molecules consist of 60 carbon atoms.

    How many such molecules are present in 12 g of this type of carbon?

    A. 1.0×10^{22}
    C. 6.0×10^{23}
    B. 1.2×10^{23}
    D. 3.6×10^{25}

12. Diabetics suffer from a deficiency of the protein insulin (relative formula mass 6000).

    What mass of insulin will contain 3×10^{20} molecules?

    A. 3 g   B. 6 g   C. 30 g   D. 60 g

In questions 13 and 14 take the molar volume to be 23 litres mol^{-1}.

13. How many molecules are in 2.3 litres of oxygen?

    A. 6×10^{22}
    C. 6×10^{23}
    B. 1.2×10^{23}
    D. 1.2×10^{24}

14. How many atoms are in 0.23 litres of hydrogen?

    A. 6×10^{21}
    C. 6×10^{22}
    B. 1.2×10^{22}
    D. 1.2×10^{23}

Test 3.5   Molar volume

For the questions in this test assume that all measurements are made at the same temperature and pressure.

1. Which gas has the smallest volume?

   A. 10 g of oxygen
   B. 10 g of carbon monoxide
   C. 10 g of ethane (C_2H_6)
   D. 10 g of hydrogen

2. Which gas has the greatest volume?

   A. 1 g of hydrogen
   B. 14 g of nitrogen
   C. 20 g of neon
   D. 35.5 g of chlorine

3. Which gas has the same volume as 1 g of helium?

   A. 1.6 g of methane
   B. 2.2 g of carbon dioxide
   C. 3.6 g of hydrogen oxide
   D. 7 g of carbon monoxide

4. Given equal volumes of each gas, in which pair do both gases have the same mass?

   A. hydrogen and helium
   B. methane and oxygen
   C. ethene and nitrogen
   D. carbon monoxide and nitrogen monoxide

5. Which gas has the highest density?

   A. CO   B. NO   C. N_2 D. C_2H_4

6. The volume of 1 g of hydrogen is 11.4 litres.

   What is the volume, in litres, of 2 mol of hydrogen?

   A. 5.7
   B. 11.4
   C. 22.8
   D. 45.6

7. A gaseous hydrocarbon has a density of 1.25 g l^{-1}. The molar volume of the gas is 22.2 litres.

   What is the molecular formula?

   A. CH_4 B. C_2H_4 C. C_3H_6 D. C_4H_8

8. The density of chlorine gas is found to be 3.00 g l^{-1}.

   Under these conditions, the molar volume, in litres is

   A. 11.8   B. 22.4   C. 23.7   D. 35.5.

9. Using the density quoted in the Data Booklet, what is the number of moles of nitrogen molecules in a 5 litre container?

   A. 0.11   B. 0.23   C. 0.35   D. 0.47

10. What is the volume, in litres, of 2.02 g of neon?

    Take the molar volume to be 23 litres mol^{-1}.

    A. 1   B. 2   C. 2.3   D. 23

Test 3.6 Calculations based on equations (i)

1. $2CO\,(g) + O_2\,(g) \rightarrow 2CO_2\,(g)$

   What mass, in grams, of carbon dioxide would be obtained by the combustion of 28 g carbon monoxide?

   A. 28 B. 44 C. 56 D. 88

2. $CH_4\,(g) + 2O_2\,(g) \rightarrow CO_2\,(g) + 2H_2O\,(l)$

   What mass, in grams, of methane is required to produce 1.8 g of water?

   A. 0.8 B. 1.6 C. 8.0 D. 16.0

3. When 2 g hydrogen is exploded in excess oxygen, how many moles of steam are produced?

   A. 0.5 B. 1 C. 2 D. 3

4. Hydrochloric acid reacts with magnesium.

   $Mg\,(s) + 2H^+\,(aq) \rightarrow Mg^{2+}(aq) + H_2\,(g)$

   What is the minimum volume of acid, concentration 4 mol l^{-1}, required to react with 0.1 mol of metal?

   A. 25 cm^3 B. 50 cm^3 C. 100 cm^3 D. 200 cm^3

5. How many moles of magnesium react with 20 cm^3 of 2 mol l^{-1} sulphuric acid?

   A. 0.01 B. 0.02 C. 0.04 D. 0.20

6. Calcium carbonate can be decomposed by heating.

   How many moles of carbon dioxide would be produced by the complete decomposition of 1 mol of calcium carbonate?

   A. 1.0 B. 0.5 C. 2.0
   D. It is impossible to say without knowing the temperature and pressure.

7. 1 mol of an alkane required 8 mol of oxygen for complete combustion. Which of the following is the formula for the alkane?

   A. C_3H_8 B. C_4H_{10} C. C_5H_{12} D. C_6H_{14}

8. Which alcohol will give 7 mol of carbon dioxide when 1 mol of it is completely burned?

   A. $CH_3CH_2CH(OH)CH(CH_3)_2$
   B. $(CH_3)_3CCH_2OH$
   C. $CH_3CH_2CH_2CH(OH)CH_2CH_2CH_3$
   D. $(CH_3)_2CHCH(OH)C(CH_3)_3$

9. If 1 mol of equally fine granules of three metals reacted with equal volumes of excess hydrochloric acid, which one should give off the most hydrogen?

   A. aluminium   B. magnesium
   C. lithium   D. They should all give same.

10. The equations represent a reaction between nitric acid and copper.

    NO_3^- (aq) + $4H^+$ (aq) + $3e^-$ → NO (g) + $2H_2O$ (l)

    Cu (s) → Cu^{2+} (aq) + $2e^-$

    How many moles of NO_3^- (aq) are required to oxidise 63.5 g of copper?

    A. 2/3   B. 1   C. 3/2   D. 2

In questions 11 to 18 take the molar volume of the gases to be 23.0 litres mol l^{-1}.

11. How many litres of oxygen are needed to react completely with 1 mol of calcium?

    A. 5.75   B. 11.5   C. 23.0   D. 46.0

12. In the reaction 2C (s) + O_2 (g) → 2CO (g) what mass of carbon, in grams, will be used to form 23.0 litres of CO?

    A. 0.6   B. 1.2   C. 6.0   D. 12.0

13. Chlorine can be produced according to the equation:

    MnO$_2$ (s) + 4HCl (aq) → MnCl$_2$ (aq) + 2H$_2$O (l) + Cl$_2$ (g)

    What volume of chlorine, in litres, will be produced by the reaction of 1 mol of hydrochloric acid?

    A. 5.75   B. 11.5   C. 23.0   D. 92.0

14. How many litres of hydrogen are needed to reduce 1 mol of iron(III) oxide completely to the metal?

    A. 11.5   B. 23.0   C. 46.0   D. 69.0

15. Potassium chlorate (KClO$_3$) can decompose on heating to give potassium chloride and oxygen.

    What volume of oxygen, in litres, would be produced by the complete decomposition of 1 mol of potassium chlorate?

    A. 11.5   B. 23.0   C. 34.5   D. 46

16. When excess of a carbonate was treated with acid, 11.5 litres of carbon dioxide was evolved.

    How many moles of hydrogen ions reacted?

    A. 0.25   B. 0.5   C. 1.0   D. 2.0

17. Potassium nitrate decomposes on heating to give potassium nitrite and oxygen.

    KNO$_3$ (s) → KNO$_2$ (s) + ½O$_2$ (g)

    What volume of oxygen, in litres, would be produced by the decomposition of 10.11 g of potassium nitrate?

    A. 1.15   B. 2.3   C. 3.45   D. 4.6

18. What volume of hydrogen, in litres, is produced when excess magnesium reacts with 100 cm^3 of dilute hydrochloric acid, concentration 1 mol l^{-1}?

    A. 1.15   B. 2.3   C. 11.5   D. 23

Test 3.7

The idea of excess

Questions 1 to 5 refer to reactions of metals.

For each of the reactions decide whether

A. reactant 1 is in excess B. reactant 2 is in excess
C. neither is in excess.

1. Reactant 1: 2.43 g of magnesium
   Reactant 2: 100 cm^3 of hydrochloric acid, concentration 0.2 mol l^{-1}

2. Reactant 1: 0.654 g of zinc
   Reactant 2: 25 cm^3 of copper(II) sulphate solution, concentration 1 mol l^{-1}

3. Reactant 1: 2.43 g of magnesium
   Reactant 2: 50 cm^3 of sulphuric acid, concentration 2 mol l^{-1}

4. Reactant 1: 6.35 g of copper
   Reactant 2: 100 cm^3 of silver nitrate solution, concentration 2 mol l^{-1}

5. Reactant 1: 2.43 g of magnesium
   Reactant 2: 100 cm^3 of hydrochloric acid, concentration 1 mol l^{-1}

6. 0.243 g of magnesium is added to 100 cm^3 of hydrochloric acid, concentration 1 mol l^{-1}.

   How much hydrogen, in moles, is produced?

   A. 0.01   B. 0.02   C. 0.1   D. 0.2

7. 5 g copper powder is added to silver nitrate solution. After some time the powder remaining is filtered off, washed with water and dried.

   The mass of the powder will be

   A. more than 5 g   B. equal to 5 g
   C. less than 5 g   D. unable to be calculated.

Chemistry in Society

8. Copper carbonate is produced in the reaction of solutions of copper(II) sulphate and sodium carbonate, both of the same concentration.

$$Na_2CO_3(aq) + CuSO_4(aq) \rightarrow CuCO_3(s) + Na_2SO_4(aq)$$

Which mixture would give the greatest mass of precipitate?

A. 1.5 cm^3 of $Na_2CO_3(aq)$ + 0.5 cm^3 of $CuSO_4(aq)$
B. 0.5 cm^3 of $Na_2CO_3(aq)$ + 1.5 cm^3 of $CuSO_4(aq)$
C. 1.0 cm^3 of $Na_2CO_3(aq)$ + 1.0 cm^3 of $CuSO_4(aq)$
D. 2.0 cm^3 of $Na_2CO_3(aq)$ + 0.5 cm^3 of $CuSO_4(aq)$

9. How much hydrogen would be released by placing 6.54 g of zinc in 200 cm^3 of hydrochloric acid, concentration 1 mol l^{-1}?

A. 0.2 mol
B. just over 0.2 mol
C. 0.1 mol
D. just over 0.1 mol

Questions 10 to 13 refer to the addition of 2.43 g of magnesium to 250 cm^3 of copper(II) sulphate solution, concentration 2 mol l^{-1}.

Decide whether each of the statements is

A. TRUE
B. FALSE.

10. All of the magnesium reacts.
12. 6.35 g of copper is displaced.

11. 0.025 mol of copper ions react.
13. 0.25 mol of magnesium reacts.

Questions 14 to 17 refer to the addition of 63.5 g of copper to 1 litre of silver nitrate solution, concentration 1 mol l^{-1}.

Decide whether each of the statements is

A. TRUE
B. FALSE.

14. The resulting solution is colourless.
16. 63.5 g of silver is formed.

15. All the copper dissolves.
17. 1 mol of silver is formed.

Test 3.8 Calculations based on equations (ii)

For the questions in this test assume that all measurements are made at the same temperature and pressure.

Questions 1 to 6 refer to reactions involving gases.

Decide whether the total volume of products for each reaction will be

A. less than the total volume of reactants
B. equal to the total volume of reactants
C. greater than the total volume of reactants.

1. $2NH_3 (g) \rightarrow N_2 (g) + 3H_2 (g)$

2. $H_2 (g) + Cl_2 (g) \rightarrow 2HCl (g)$

3. $N_2 (g) + 2O_2 (g) \rightarrow 2NO_2 (g)$

4. $C (s) + O_2 (g) \rightarrow CO_2 (g)$

5. $C_2H_4 (g) + 3O_2 (g) \rightarrow 2CO_2 (g) + 2H_2O (l)$

6. $Fe_2O_3 (s) + 3CO (g) \rightarrow 2Fe (s) + 3CO_2 (g)$

7. The reaction of hydrogen and oxygen is represented by the equation:

    $H_2(g) + ½O_2(g) \rightarrow H_2O(g)$

    What is the volume of oxygen, in litres, which reacts with 1 litre of hydrogen?

    A. 1/4   B. 1/2   C. 1   D. 2

8. What volume of oxygen, in litres, is required for the complete combustion of 1 litre of butane?

    A. 1   B. 4   C. 6.5   D. 13

Chemistry in Society

9. How many litres of carbon dioxide would be obtained by the complete combustion of 2 litres of ethene?

   A. 2 B. 4 C. 6 D. 8

10. What volume of oxygen, in litres, would be required for the complete combustion of a gaseous mixture containing 1 litre of carbon monoxide and 3 litres of hydrogen?

    A. 1 B. 2 C. 3 D. 4

11. $N_2 (g) + 2O_2 (g) \rightarrow 2NO_2 (g)$

    How many litres of nitrogen dioxide gas could be obtained by sparking 5 litres of nitrogen gas with 2 litres of oxygen gas?

    A. 2 B. 3 C. 4 D. 5

12. $2NO (g) + O_2 (g) \rightarrow 2NO_2 (g)$

    How many litres of nitrogen dioxide gas could be obtained by mixing 3 litres of nitrogen monoxide gas and 1 litre of oxygen gas?

    A. 2 B. 3 C. 4 D. 5

13. A mixture of 8 litres of oxygen and 12 litres of hydrogen is sparked at 200 °C.

    What is the total volume of the gas, in litres, at the end of the reaction?

    A. 8 B. 12 C. 14 D. 20

14. What volume of carbon dioxide, in cm^3, would be obtained by the combustion of 28 cm^3 of carbon monoxide?

    A. 28 B. 42 C. 56 D. 84

15. A volume of 10 cm^3 of carbon monoxide was passed over heated copper(II) oxide until no further reaction occured.

    What volume of gas, in cm^3, was obtained?

    A. 0 B. 10 C. 15 D. 20

16. The composition of air by volume is approximately 20% oxygen, 80% nitrogen.

    When air is passed through red-hot carbon, the following reaction occurs:

    2C (s)   + O_2 (g)   →   2CO (g)

    If all the oxygen is converted to carbon monoxide, what is the composition, by volume, of the gas produced?

    A. 20% carbon monoxide, 80% nitrogen
    B. 33% carbon monoxide, 66% nitrogen
    C. 40% carbon monoxide, 60% nitrogen
    D. 50% carbon monoxide, 50% nitrogen

Questions 17 and 18 refer to the following reaction.

3CuO (s)   +   2NH_3 (g)   →   3Cu(s)   + N_2 (g)   +   3H_2O (g)

The reaction is carried out in a furnace heated to 150 °C.

17. What volume of gas, in cm^3, would be obtained by the reaction between 100 cm^3 of ammonia gas and excess copper(II) oxide?

    A. 50   B. 100   C. 200   D. 400

18. The mixture is then allowed to cool to room temperature.

    What volume of gas, in cm^3, would remain?

    A. 50   B. 100   C. 150   D. 200

19. 20 cm^3 of butane is burned in 150 cm^3 of oxygen and the reaction mixture allowed to cool to room temperature.

    C_4H_{10} (g)   +   6½O_2 (g)   →   4CO_2 (g)   +   5½H_2O (l)

    What is the volume, in cm^3, of the resulting gas mixture?

    A. 80   B. 100   C. 180   D. 200

Chemistry in Society

20. 50 cm³ of propane is mixed with 500 cm³ of oxygen and the mixture is ignited. The products are allowed to cool to room temperature.

$C_3H_8(g)$ + $5O_2(g)$ → $3CO_2(g)$ + $4H_2O(l)$

What is the volume, in cm³, of the resulting gas mixture?

A. 150   B. 300   C. 400   D. 700

Questions 21 and 22 refer to the following experiment.

A mixture of 50 cm³ of carbon monoxide and 40 cm³ carbon dioxide is heated with excess copper(II) oxide until no further reaction occurs.

21. What is the total volume of gas, in cm³, after the reaction?

A. 40   B. 50   C. 90   B 140

22. If the gases are then passed through an aqueous solution of sodium hydroxide, what is the volume, in cm³, of the remaining gas?

A. 40   B. 50   C. 90   D. 140

Questions 23 and 24 refer to the combustion of methane.

15 cm³ of methane was collected in a tube over mercury and 35 cm³ of oxygen was added. The mixture was then sparked to burn the methane and the reaction mixture allowed to cool to room temperature.

$CH_4(g)$ + $2O_2(g)$ → $CO_2(g)$ + $2H_2O(l)$

23. What was the volume, in cm³, of the remaining gas?

A. 20   B. 45   C. 50   D. 55

24. If a small volume of sodium hydroxide solution was injected into the tube, what would be the volume, in cm³, of the remaining gas?

A. 5   B. 15   C. 30   D. 45

Test 3.9

Equilibrium (i)

The questions in this test refer to reversible reactions at equilibrium.

Decide whether each of the statements is

A. TRUE **B.** FALSE.

1. The concentrations of reactants are always equal to the concentrations of products.

2. The concentrations of reactants and products are constant.

3. Molecules of reactants are no longer changing into molecules of products.

4. The rates of forward and reverse reactions are equal.

5. The activation energies of the forward and reverse reactions are equal.

6. Catalysts decrease the time required for the equilibrium to be established.

7. Catalysts alter the position of equilibrium.

8. Catalysts lower the activation energy of the forward reactions.

9. Catalysts increase the rate of the reverse reactions.

10. Catalysts increase the activation energy of the reverse reactions.

11. Catalysts increase the enthalpy change for the forward reaction.

Decide whether each of the following

A. influences the position of equilibrium
B. does **NOT** influence the position of equilibrium.

12. particle size
13. reactant concentration
14. catalytic action
15. temperature change

Chemistry in Society

16. Which line in the table shows the effect of a catalyst on the reaction rates and position of equilibrium in a reversible reaction?

	Rate of forward reaction	Rate of reverse reaction	Position of equilibrium
A.	increased	increased	moved to right
B.	increased	increased	unchanged
C.	increased	decreased	moved to right
D.	increased	decreased	unchanged

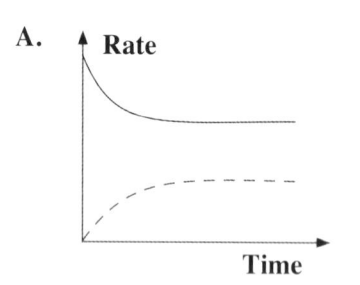

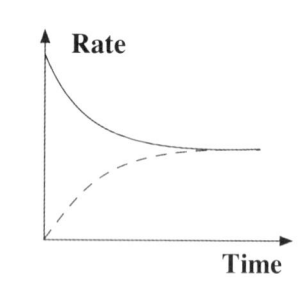

17. Which graph shows how the rates of the forward and reverse reactions change when reactants and products form an equilibrium mixture.

Key: ——— forward reaction
     — — — reverse reaction

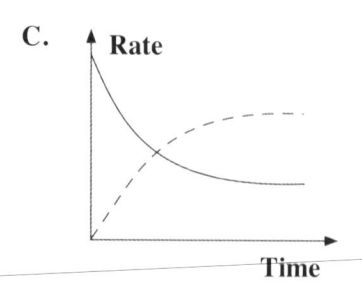

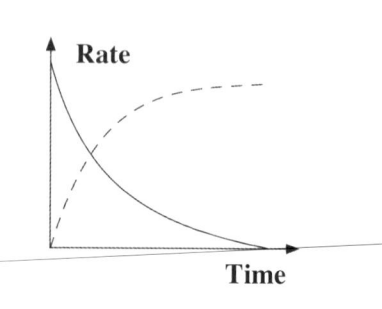

18. Which graph **cannot** be used to show changes in the concentration of reactants and products as an equilibrium mixture is formed.

Key: —— concentration of reactants
 — — — concentration of products

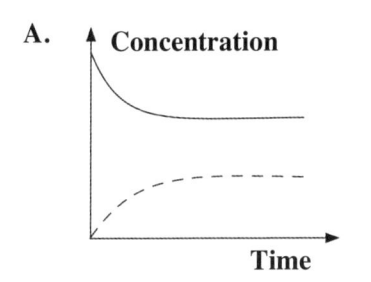

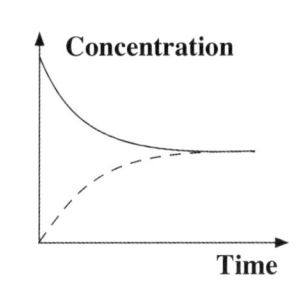

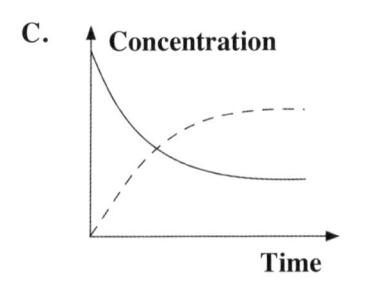

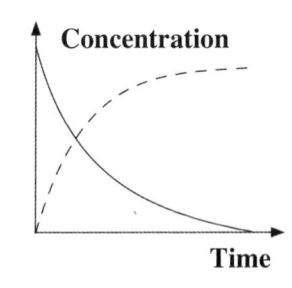

A. Concentration vs Time

B. Concentration vs Time

C. Concentration vs Time

D. Concentration vs Time

Test 3.10

Equilibrium (ii)

Questions 1 to 3 refer to the equilibrium:

$$N_2 (g) + 3H_2 (g) \rightleftharpoons 2NH_3 (g)$$

Changing the concentration of reactants and products can

- A. move the equilibrium to the right
- B. move the equilibrium to the left.

What is the effect on the equilibrium mixture of each of the following changes?

1. increasing the concentration of nitrogen gas

2. decreasing the concentration of hydrogen gas

3. decreasing the concentration of ammonia gas

Questions 4 to 11 refer to the equilibrium:

$$Cl_2 (aq) + H_2O (g) \rightleftharpoons 2H^+ (aq) + ClO^- (aq) + Cl^- (aq)$$

The addition of substances can

- A. move the equilibrium to the product side
- B. move the equilibrium to the reactant side
- C. leave the equilibrium mixture unchanged.

What is the effect of adding each of the following substances?

4. sodium chloride crystals

5. nitric acid

6. potassium sulphate crystals

7. silver nitrate solution

8. sodium hydroxide solution

9. potassium nitrate solution

10. hydrogen chloride

11. hydrogen

Questions 12 to 18 refer to the effect of an increase in pressure on chemical reactions at equilibrium.

An increase in pressure can

    A.   increase the concentration of reactants
    B.   increase the concentration of products
    C.   have no effect on the concentration of reactants and products.

What is the effect of an increase in pressure in each of the following reactions?

12. $N_2O_4(g) \rightleftharpoons 2NO_2(g)$

13. $H_2(g) + I_2(g) \rightleftharpoons 2HI(g)$

14. $2SO_2(g) + O_2(g) \rightleftharpoons 2SO_3(g)$

15. $C(s) + H_2O(g) \rightleftharpoons H_2(g) + CO(g)$

16. $CO(g) + H_2O(g) \rightleftharpoons CO_2(g) + H_2(g)$

17. $NH_3(g) + H_2O(g) \rightleftharpoons NH_4^+(aq) + OH^-(aq)$

18. $Fe_2O_3(s) + 3CO(g) \rightleftharpoons 2Fe(s) + 3CO_2(g)$

Questions 19 to 22 refer to the effect of a change in temperature on chemical reactions at equilibrium.

What will be the effect of the temperature change in each of the following reactions?

    A.   increase the concentration of reactants
    B.   increase the concentration of products

19. temperature increase
$PCl_5(g) \rightleftharpoons PCl_3(g) + Cl_2(g)$ $\Delta H = -92$ kJ mol^{-1}

20. temperature decrease
$2NO(g) \rightleftharpoons N_2(g) + O_2(g)$ $\Delta H = -180$ kJ mol^{-1}

21. temperature increase
$H_2O(g) \rightleftharpoons 2H_2(g) + O_2(g)$ $\Delta H = +484$ kJ mol^{-1}

22. temperature decrease
$KBr(s) + (aq) \rightleftharpoons K^+(aq) + Br^-(aq)$ $\Delta H = +20$ kJ mol^{-1}

Test 3.11   Equilibrium (iii)

1. 0.1 mol of methanol, 0.1 mol of ethanoic acid and a few drops of concentrated sulphuric acid were warmed together. After a considerable time the reaction mixture was found still to contain some of each of the reactants as well as some ester.

   What is the best explanation of the incomplete reaction?

   A. An equilibrium mixture was formed.
   B. The temperature was too low.
   C. Insufficient methanol was used.
   D. Insufficient catalyst was used.

2. $C_2H_4(g) + H_2(g) \rightarrow C_2H_6(g)$ ΔH is -ve

   Which procedure would **not** affect the position of equilibrium?

   A. decreasing the pressure
   B. decreasing the temperature
   C. adding a catalyst
   D. adding more hydrogen

3. $2SO_2(g) + O_2(g) \rightleftharpoons 2SO_3(g)$

   In the presence of a catalyst the equilibrium yield would be

   A. increased and attained more rapidly
   B. increased and attained in the same time
   C. unchanged but attained more rapidly
   D. decreased but attained more rapidly.

4. $Ag^+(aq) + Fe^{2+}(aq) \rightleftharpoons Ag(s) + Fe^{3+}(aq)$

   Which compound when added to the equilibrium mixture would lead to an increase in the mass of silver deposited?

   You may wish to use the Data Booklet.

   A. iron(II) sulphate   B. iron(III) sulphate
   C. iron(II) hydroxide   D. iron(III) hydroxide

Questions 5 and 8 refer to the most favourable conditions for reactions.

    A.   high temperature, high pressure
    B.   high temperature, low pressure
    C.   low temperature, high pressure
    D.   low temperature, low pressure

5. $2NO(g) + O_2(g) \rightleftharpoons 2NO_2(g)$ $\Delta H = -560$ kJ mol^{-1}

   Which conditions favour the formation of NO_2?

6. $CH_4(g) + H_2O(g) \rightleftharpoons CO(g) + 3H_2(g)$ $\Delta H = +206$ kJ mol^{-1}

   Which conditions favour the formation of hydrogen?

7. $3O_2(g) \rightleftharpoons 2O_3(g)$ $\Delta H = +286$ kJ mol^{-1}

   Which conditions favour the formation of $O_3(g)$?

8. $ICl_3(l) + Cl_2(g) \rightleftharpoons ICl_5(s)$ $\Delta H = -106$ kJ mol^{-1}

   Which conditions favour the formation of $ICl_5(s)$?

9. Excess sodium chloride was shaken with water, giving a saturated solution with some solid sodium chloride on the bottom of the container.

   $NaCl(s) \rightleftharpoons Na^+(aq) + Cl^-(aq)$

   What will happen if HCl(g) is passed through the solution?

    A.   Chlorine gas will form.
    B.   The pH will rise.
    C.   Some sodium chloride will crystallise out.
    D.   Some solid sodium chloride will dissolve.

10. The decomposition of magnesium carbonate by heat can be prevented from going to completion by

    A.   absorbing the carbon dioxide produced in lime water
    B.   removing magnesium oxide as it is formed
    C.   carrying out the reaction in a small, closed flask
    D.   reducing the pressure in the reaction flask.

11. Two flasks, **X** and **Y**, with their contents as shown, were placed in a vessel of water at 40 °C.

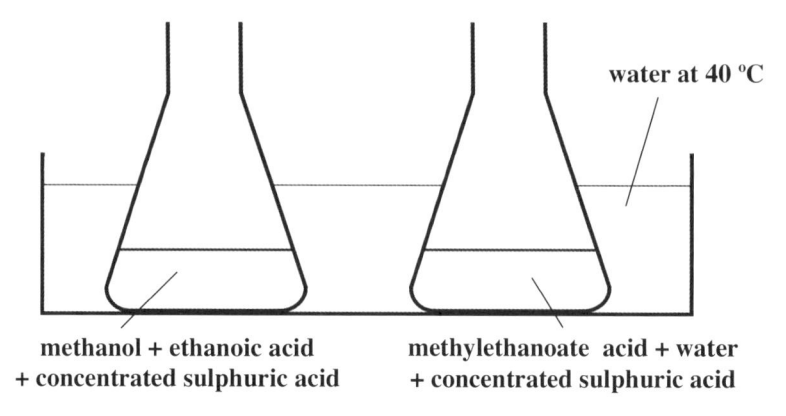

methanol + ethanoic acid + concentrated sulphuric acid

methylethanoate acid + water + concentrated sulphuric acid

water at 40 °C

After several hours the contents of both flasks were analysed.

Which of the following would be expected?

A. Flask **X** contains methyl ethanoate, methanol and ethanoic acid; flask **Y** is unchanged.
B. Flask **X** and flask **Y** both contain methyl ethanoate, methanol and ethanoic acid.
C. Flask **X** contains methyl ethanoate; flask **Y** is unchanged.
D. Flask **X** contains methyl ethanoate; flask **Y** contains methyl ethanoate, methanol and ethanoic acid.

12. If ammonia is added to a solution containing copper(II) ions, an equilibrium is set up.

$$Cu^{2+} (aq) + 2OH^- (aq) + 4NH_3 (aq) \rightleftharpoons Cu(NH_3)_4(OH)_2 (aq)$$
(deep blue)

If acid is added to this equilibrium mixture

A. the intensity of the deep blue colour will increase
B. the equilibrium position will move to the right
C. the concentration of Cu^{2+} (aq) ions will increase
D. the equilibrium position will **not** be affected.

13. Gaseous iodine and hydrogen react to form an equilibrium mixture.

    $I_2(g) + H_2(g) \rightleftharpoons 2HI(g)$

    Which graph shows the pressure inside a sealed container as the reaction proceeds at a constant temperature?

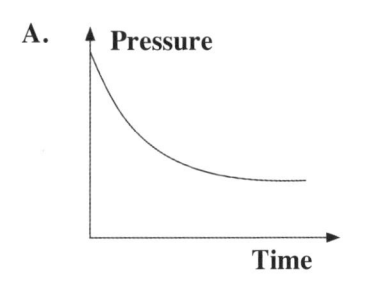

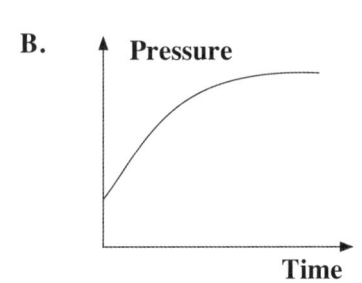

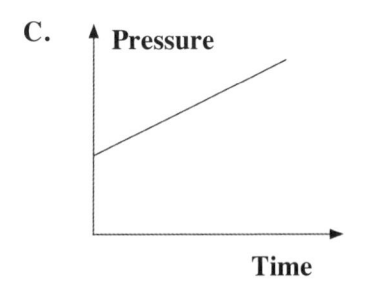

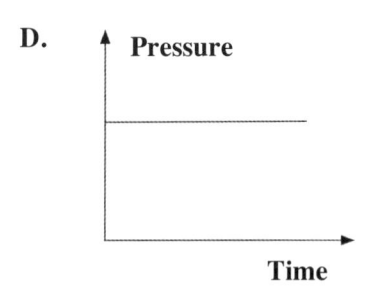

    A. [Pressure vs Time graph: decreasing curve levelling off]

    B. [Pressure vs Time graph: increasing curve levelling off]

    C. [Pressure vs Time graph: straight line increasing]

    D. [Pressure vs Time graph: horizontal line]

14. The following equilibrium exists in bromine water.

    $Br_2(aq) + H_2O(g) \rightleftharpoons Br^-(aq) + 2H^+(aq) + OBr^-(aq)$
    (red)   (colourless)   (colourless)

    The red colour of bromine would fade on adding a few drops of a concentrated solution of

    A. HCl
    B. KBr
    C. AgNO$_3$
    D. NaOBr.

Chemistry in Society

Test 3.12

Percentage yield

Calculate the percentage yield for each reaction.

1. $N_2 (g)$ + $3H_2 (g)$ → $2NH_3 (g)$

   Under test conditions, 60 g of hydrogen reacts with excess nitrogen to produce 80 g of ammonia.

   A. 80/340 x 100
   B. 80/170 x 100
   C. 30/80 x 100
   D. 60/80 x 100

2. $CH_3CH(OH)CH_3$ → CH_3COCH_3

   In a preparation, 6.4 g of propanone is obtained from 8.0 g of propan-2-ol.

   A. 83   B. 86   C. 89   D. 92

3. CH_3CH_2OH + $HCOOH$ → $HCOOCH_2CH_3$

   In a preparation, 37.2 g of ethyl methanoate is obtained from 28.3 g of ethanol.

   A. 76   B. 79   C. 82   D. 85

4. $H_2 (g)$ + $I_2 (g)$ → $2HI (g)$

   Under test conditions, 20 g of hydrogen produce 1800 g of hydrogen iodide

   A. 17.5   B. 35   C. 52.5   D. 70

5. $2SO_2 (g)$ + $O_2 (g)$ → $2SO_3 (g)$

   Under test conditions, 2 kg of sulphur dioxide reacts with excess oxygen to produce 0.5 kg of sulphur trioxide.

   A. 20   B. 25   C. 30   D. 40

6. $Fe_2O_3 (s)$ + $3CO (g)$ → $2Fe (s)$ + $3CO_2 (g)$

   Under test conditions, 2.0 tonnes of iron is obtained from 3.2 tonnes of iron(III) oxide.

   A. 45   B. 80   C. 89   D. 95

Test 3.13

Atom economy

Calculate the atom economy for each reaction.

1. Making ethene

   C_2H_6 → C_2H_4 + H_2

   A. 63.3%   **B.** 73.3%   **C.** 83.3%   **D.** 93.3%

2. Making sulphur trioxide

   $2SO_2$ + O_2 → $2SO_3$

   A. 70%   **B.** 80%   **C.** 90%   **D.** 100%

3. Making hydrogen

   C + H_2O → H_2 + CO

   A. 6.7%   **B.** 9.2%   **C.** 11.7%   **D.** 14.2%

4. Making carbon dioxide

   $2NaHCO_3$ + H_2SO_4 → $2CO_2$ + Na_2SO_4 + $2H_2O$

   A. 16.5%   **B.** 26.2%   **C.** 33.1%   **D.** 52.3%

5. Making ammonia

   $CO(NH_2)_2$ + H_2O → $2NH_3$ + CO_2

   A. 21.8%   **B.** 28.3%   **C.** 43.6%   **D.** 56.6%

6. Making chlorine

   $Ca(OCl)_2$ + 2HCl → $Ca(OH)_2$ + $2Cl_2$

   A. 24.8%   **B.** 32.9%   **C.** 49.6%   **D.** 65.7%

7. Making methane

   Al_4C_3 + $12H_2O$ → $3CH_4$ + $4Al(OH)_3$

   A. 4.4%   **B.** 13.3%   **C.** 24.4%   **D.** 33.3%

Test 3.14 — Enthalpy of combustion

1. The sign for an enthalpy of combustion
   - A. is always positive
   - B. is always negative
   - C. can be either positive or negative.

2. Which equation represents an enthalpy of combustion?
   - A. C_2H_6 (g) + $3O_2$ (g) → $2CO_2$ (g) + $3H_2O$ (l)
   - B. C_2H_5OH (l) + O_2 (g) → CH_3COOH(g) + H_2O (l)
   - C. CH_2CHO (l) + O_2 (g) → CH_3COOH (l)
   - D. CH_4 (g) + $1½O_2$ (g) → CO(g) + $2H_2O$ (l)

3. Ethanol (C_2H_5OH) has a different enthalpy of combustion from dimethyl ether (CH_3OCH_3). This is mainly because the compounds have different
   - A. molecular masses
   - B. bonds within the molecules
   - C. products of combustion
   - D. boiling points.

4. When 1 g of an alcohol (formula mass 46) is burned, 30 kJ of energy is released.

   What is the enthalpy of combustion, in kJ mol^{-1}, of the alcohol?
   - A. -30
   - B. -1380
   - C. -650
   - D. -1920.

For questions 5 to 7, use the Enthalpies of Combustion given in the Data Booklet.

5. What is the approximate enthalpy of combustion of butan-1-ol, in kJ mol^{-1}?
   - A. -2237
   - B. -2453
   - C. 2685
   - D. 2890

6. What is the enthalpy change, in kJ, when 3.2 g of methanol is burned?
   - A. -72.6
   - B. +72.6
   - C. -726
   - D. +726

7. What is the mass of ethanol, in grams, which has to be burned to produce 13.67 kJ?
   - A. 0.46
   - B. 4.6
   - C. 13.67
   - D. 1367

8. A hydrocarbon burned to raise the temperature of 100 cm³ of water by 23 °C.

   Approximately how much energy, in kJ mol⁻¹, is produced?

   A.   9.6   B.   96   C.   960   D.   9600

9. When 2.24 litres of a gas was burned, the heat produced warmed 2 litres of water from 12 °C to 38 °C.

   What is the approximate enthalpy of combustion, in kJ mol⁻¹, of the gas?

   (Take the volume of one mole of gas to be 22.4 litres.)

   A.   -21.7   B.   -217   C.   -2170   D.   -21 700

10. When 1.5 g of a gas (formula mass 30) is burned, the heat produced raised the temperature of 500 cm³ of water by 37 °C.

    What is the approximate enthalpy of combustion, in kJ mol⁻¹, of the gas?

    A.   -77   B.   -155   C.   -775   D.   -1550

Test 3.15 — Hess's Law

1. Given the equations:

Mg (s)	+	2H$^+$ (aq)	→	Mg^{2+} (aq) + H$_2$ (g)		ΔH = a
Zn (s)	+	2H$^+$ (aq)	→	Zn^{2+} (aq) + H$_2$ (g)		ΔH = b
Mg (s)	+	Zn^{2+} (aq)	→	Mg^{2+} (aq) + Zn (s)		ΔH = c

   then according to Hess's Law

   A. a + b = -c
   B. a + b = c
   C. a + c = b
   D. a - b = c.

2. What is the relationship between **a, b, c** and **d**?

S (s)	+	H$_2$ (g)	→	H$_2$S (g)	ΔH = a
H$_2$ (g)	+	½O$_2$ (g)	→	H$_2$O (l)	ΔH = b
S (s)	+	O$_2$ (g)	→	SO$_2$ (g)	ΔH = c
H$_2$S (g)	+	1½O$_2$ (g)	→	H$_2$O (l) + SO$_2$ (g)	ΔH = d

   A. a = b + c - d
   B. a = d - b - c
   C. a = b - c - d
   D. a = d + c - b

3. The enthalpies of combustion of C(s), H$_2$ (g) and C$_4$H$_9$OH (l) are:

C (s)	+	O$_2$ (g)	→	CO$_2$ (g)	ΔH = a
H$_2$ (g)	+	½O$_2$ (g)	→	H$_2$O (l)	ΔH = b
C$_4$H$_9$OH (l)	+	6O$_2$ (g)	→	4CO$_2$ (g) + 5 H$_2$O (l)	ΔH = c

   The enthalpy of formation of butanol is the enthalpy change for the reaction:

   4C (s) + 5H$_2$ (g) + ½O$_2$ (g) → C$_4$H$_9$OH (l)

   What is the enthalpy of formation of butanol?

   A. 4a + 5b - c
   B. 2a + 10b - c
   C. c - 4a - 5b
   D. 2a + 5b + c

4. Given that the enthalpies of combustion of carbon, hydrogen and ethane are **X**, **Y** and **Z** respectively, the enthalpy change for the reaction

   2C (s) + 3H$_2$ (g) → C$_2$H$_6$ (g)

   will be

   A. 2X + 3Y - Z
   B. 2X + 3Y + Z
   C. X + Y - Z
   D. -2X - 3Y + Z.

5. The enthalpies of combustion for propene and propane are ΔH_1 and ΔH_2 respectively.

   Given the equations

   3C (s) + 3H$_2$ (g) → C$_3$H$_6$ (g) ΔH_3
   3C (s) + 4H$_2$ (g) → C$_3$H$_8$ (g) ΔH_4

   the enthalpy change for the reaction

   C$_3$H$_6$ (g) + H$_2$ (g) → C$_3$H$_8$ (g)   is

   A. $\Delta H_1 - \Delta H_2$
   B. $\Delta H_2 - \Delta H_1$
   C. $\Delta H_3 - \Delta H_4$
   D. $\Delta H_4 - \Delta H_3$.

6. C (s) + O$_2$ (g) → CO$_2$ (g) ΔH = -394 kJ mol^{-1}

   CO (g) + ½O$_2$ (g) → CO$_2$ (g) ΔH = -282 kJ mol^{-1}

   The enthalpy change for the reaction

   C (s) + ½O$_2$ (g) → CO (g)

   will be

   A. +112   B. -676   C. -112   D. +676

Chemistry in Society   127

7. Consider the reaction pathway shown.

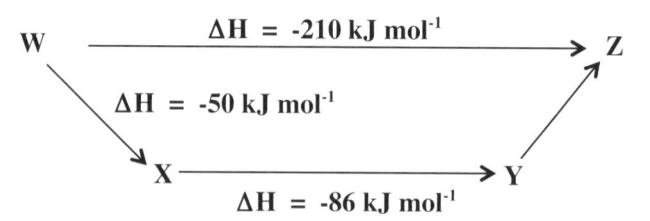

W $\xrightarrow{\Delta H = -210 \text{ kJ mol}^{-1}}$ Z

W $\xrightarrow{\Delta H = -50 \text{ kJ mol}^{-1}}$ X $\xrightarrow{\Delta H = -86 \text{ kJ mol}^{-1}}$ Y $\rightarrow$ Z

According to Hess's Law, the ΔH value, in kJ mol⁻¹, for reaction Z to Y is

A. +74 B. -74 C. +346 D. -346.

8. $N_2(g)$ + $2O_2(g)$ → $2NO_2(g)$ $\Delta H = +88 \text{ kJ mol}^{-1}$
    $N_2(g)$ + $2O_2(g)$ → $N_2O_4(g)$ $\Delta H = +10 \text{ kJ mol}^{-1}$

The enthalpy change, in kJ mol⁻¹, for the reaction

$2NO_2(g)$ → $N_2O_4(g)$

will be

A. +98 B. +78 C. -78 D. -98.

9. 2Fe (s) + 1½O_2(g) → Fe_2O_3(s) $\Delta H = -827 \text{ kJ mol}^{-1}$
    2Al (s) + 1½O_2(g) → Al_2O_3(s) $\Delta H = -1676 \text{ kJ mol}^{-1}$

The enthalpy change, in kJ mol⁻¹, for the reaction

Fe_2O_3(s) + 2Al (s) → 2Fe (s) + Al_2O_3(s)

is given by

A. -827 + 1676
B. 2(-827 + 1676)
C. +827 - 1676
D. ½(827 - 1676).

10. Consider the reaction pathway shown.

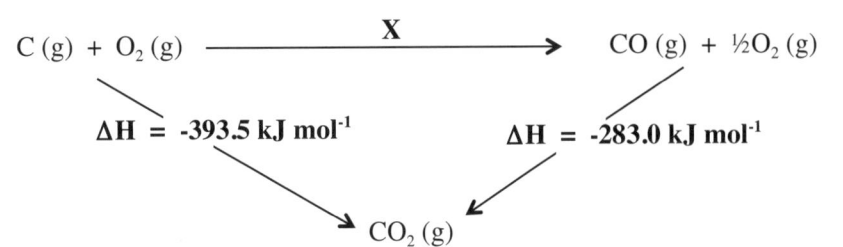

$$C(g) + O_2(g) \xrightarrow{X} CO(g) + \tfrac{1}{2}O_2(g)$$

$\Delta H = -393.5$ kJ mol^{-1} (C + O$_2$ → CO$_2$)

$\Delta H = -283.0$ kJ mol^{-1} (CO + ½O$_2$ → CO$_2$)

→ CO$_2$(g)

According to Hess's Law, the enthalpy change, in kJ mol^{-1}, for reaction **X** is

A. +110.5   B. −110.5   C. −676.5   D. +676.5.

11. $CH_2Cl_2(g) + O_2(g) \rightarrow CO_2(g) + 2HCl(g)$ $\Delta H = -446$ kJ mol^{-1}
    $C(s) + O_2(g) \rightarrow CO_2(g)$ $\Delta H = -394$ kJ mol^{-1}
    $H_2(g) + Cl_2(g) \rightarrow 2HCl(g)$ $\Delta H = -92$ kJ mol^{-1}

The enthalpy change, in kJ mol^{-1}, for the reaction

$C(s) + H_2(g) + Cl_2(g) \rightarrow CH_2Cl_2(g)$

will be

A. −932   B. −40   C. +144   D. +748.

12. $6C(s) + 5H_2(g) \rightarrow C_6H_{10}(l)$ $\Delta H = -3$ kJ mol^{-1}
    $6C(s) + 6H_2(g) \rightarrow C_6H_{12}(l)$ $\Delta H = +129$ kJ mol^{-1}

The enthalpy of hydrogenation of cyclohexene to cyclohexane, in kJ mol^{-1}, will be

A. −132   B. −129   C. +129   D. +132.

13. What is the enthalpy change, in kJ mol^{-1}, for the complete hydrogenation of one mole of ethene, C_2H_4?

Use the Enthalpies of Combustion in the Data Booklet.

A. −462   B. +1094   C. +3344   D. +3916

Chemistry in Society

Test 3.16

Bond enthalpy

In questions 1 to 6, decide whether each of the changes has a ΔH value that

A. has a positive sign
B. has a negative sign
C. represents a bond enthalpy
D. represents a mean bond enthalpy.

For each of the questions, **two** answers should be given.

1. H-H (g) → H (g) + H (g)

2. N-H (g) → N (g) + H (g)

3. O (g) + O (g) → O=O (g)

4. H (g) + Cl (g) → H-Cl (g)

5. C (g) + Cl (g) → C-Cl (g)

6. C=C (g) → C (g) + C (g)

7. The mean bond enthalpy for the N-H bond is equal to one third of the ΔH for which change?

   A. N (g) + 3H (g) → NH_3 (g)
   B. N_2 (g) + $3H_2$ (g) → $2NH_3$ (g)
   C. $½N_2$ (g) + $1½H_2$ (g) → NH_3 (g)
   D. NH_3 (g) → $½N_2$ (g) + $1½H_2$ (g)

8. Hydrogen and chlorine react in the presence of bright light.
   One step in the reaction is shown.

   $$H_2(g) + Cl(g) \rightarrow HCl(g) + H(g)$$

   Which calculation gives a measure of the enthalpy change for this reaction?

   A. H-H bond enthalpy + Cl-Cl bond enthalpy
   B. H-H bond enthalpy − Cl-Cl bond enthalpy
   C. H-H bond enthalpy + H-Cl bond enthalpy
   D. H-H bond enthalpy − H-Cl bond enthalpy

9. The mean bond enthalpy of the C-H bond is 412 kJ mol^{-1}.

   From this information, it can be calculated that 1648 kJ of energy is

   A. evolved when 1 mol of methane is burned in excess oxygen
   B. required to dissociate 1 mol of methane into free carbon atoms and hydrogen atoms
   C. required for the complete combustion of 1 mol of methane
   D. evolved when 1 mol of graphite combines with 2 mol of hydrogen gas.

10. The mean bond enthalpy of the C-F bond is 484 kJ mol^{-1}.

    In which of the processes is ΔH approximately equal to 1936 kJ mol^{-1}?

    A. $CF_4(g) \rightarrow C(s) + 2F_2(g)$
    B. $CF_4(g) \rightarrow C(g) + 4F(g)$
    C. $CF_4(g) \rightarrow C(g) + 2F_2(g)$
    D. $CF_4(g) \rightarrow C(s) + 4F(g)$

11. The mean bond enthalpy of the C-H bond is equal to one quarter of the value of the ΔH for which change?

    A. $CH_4(g) \rightarrow C(g) + 4H(g)$
    B. $CH_4(g) \rightarrow C(s) + 4H(g)$
    C. $C(s) + 2H_2(g) \rightarrow CH_4(g)$
    D. $CH_4(g) + O_2(g) \rightarrow CO_2(g) + 2H_2O(g)$

In questions 12 to 17 calculate the enthalpy change, in kJ mol⁻¹, for each of the reactions.

Use the Selected Bond and Mean Bond Enthalpies in the Data Booklet.

12. H−H (g)   +   I−I (g) →   2 H−I (g)

    A. +9   B. -9   C. +289   D. -289

13. H−CH₂−H (g) + Cl−Cl (g) → H−CHCl−H (g) + H−Cl (g)

    (methane + Cl₂ → chloromethane + HCl)

    A. +115   B. -115   C. +189   D. -189

14. H−CH₂−CH₂−H + Br−Br → H−CH₂−CHBr−H + H−Br

    (ethane + Br₂ → bromoethane + HBr)

    A. +36   B. -36   C. +1248   D. -1248

15. H₂ (g)   +   Cl (g) →   HCl (g)   +   H (g)

    A. 4   B. 243   C. 432   D. 436

16. CH₃-CH=CH-CH₃ + H₂ (g) → CH₃-CH₂-CH₂-CH₃

    A. -244   B. +244   C. -124   D. +124

17. propene + hydrogen chloride → chloropropane

    A. -54   B. +284   C. +294   D. +358

Test 3.17   **Oxidation and reduction (revision)**

You may wish to use the Data Booklet for the questions in this test.

Decide whether each of the reactions involve

    A.   oxidation   **B.**   reduction.

1. $Mg^{2+}(aq) + 2e^- \rightarrow Mg(s)$

2. $Ag(s) \rightarrow Ag^+(aq) + e^-$

3. $2Cl^-(aq) \rightarrow Cl_2(aq) + 2e^-$

4. $Fe^{2+}(aq) + 2e^- \rightarrow Fe(s)$

5. $2I^-(aq) \rightarrow I_2(aq)$

6. $Cu(s) \rightarrow Cu^{2+}(aq)$

7. $SO_3^{2-}(aq) \rightarrow SO_4^{2-}(aq)$

8. $MnO_4^-(aq) \rightarrow Mn^{2+}(aq)$

9. iron(II) → iron(III)

10. cobalt(III) → cobalt(II)

11. zinc atoms → zinc ions

12. bromine molecules → bromide ions

Test 3.18 Oxidising and reducing agents

You may wish to use the Data Booklet for the questions in this test.

Decide whether each of the underlined reactants is acting as

    A.   an oxidising agent   **B.**   a reducing agent.

1. $\underline{Zn}\,(s)\ +\ 2H^+\,(aq)\ \rightarrow\ Zn^{2+}\,(aq)\ +\ H_2\,(g)$

2. $\underline{Cl_2}\,(aq)\ +\ 2I^-\,(aq)\ \rightarrow\ 2Cl^-\,(aq)\ +\ I_2\,(aq)$

3. $6\underline{Fe^{2+}}\,(aq)\ +\ Cr_2O_7^{2-}\,(aq)\ +\ 14H^+\,(aq)\ \rightarrow\ 6Fe^{3+}\,(aq)\ +\ 2Cr^{3+}\,(aq)\ +\ 7H_2O\,(l)$

4. $\underline{Mg}\,(s)\ +\ Cu^{2+}\,(aq)\ \rightarrow\ Mg^{2+}\,(aq)\ +\ Cu\,(s)$

5. $Fe_2O_3\,(s)\ +\ 3\underline{CO}\,(g)\ \rightarrow\ 2Fe\,(s)\ +\ 3CO_2\,(g)$

6. $Mg\,(s)\ +\ 4\underline{HNO_3}\,(aq)\ \rightarrow\ Mg(NO_3)_2\,(aq)\ +\ 2NO_2\,(g)\ +\ 2H_2O\,(l)$

7. $2\underline{Na_2S_2O_2}\,(aq)\ +\ I_2\,(aq)\ \rightarrow\ 2NaI\,(aq)\ +\ Na_2S_4O_6\,(aq)$

8. $\underline{SnCl_2}\,(aq)\ +\ HgCl_2\,(aq)\ \rightarrow\ Hg\,(l)\ +\ SnCl_4\,(aq)$

9. $Zn\,(s)\ +\ 2\underline{AgNO_3}\,(aq)\ \rightarrow\ Zn(NO_3)_2\,(aq)\ +\ 2Ag\,(s)$

10. $C_6H_8O_6\,(aq)\ +\ \underline{Br_2}\,(aq)\ \rightarrow\ C_6H_6O_6\,(aq)\ +\ 2HBr\,(aq)$

Test 3.19

Redox reactions

Decide whether each of the following

A. is a redox reaction **B.** is **NOT** a redox reaction.

1. $Mg\,(s) + 2H^+\,(aq) \rightarrow Mg^{2+}\,(aq) + H_2\,(g)$

2. $N_2\,(g) + 3H_2\,(g) \rightarrow 2NH_3\,(g)$

3. $H^+\,(aq) + OH^-\,(aq) \rightarrow H_2O\,(l)$

4. $CuO\,(s) + CO\,(g) \rightarrow Cu\,(s) + CO_2\,(g)$

5. $Br_2\,(aq) + 2I^-\,(aq) \rightarrow 2Br^-\,(aq) + I_2\,(aq)$

6. $C_2H_4\,(g) + H_2\,(g) \rightarrow C_2H_6\,(g)$

7. $Ba^{2+}\,(aq) + SO_4^{2-}\,(aq) \rightarrow BaSO_4\,(s)$

8. $2Al\,(s) + 3O_2\,(g) \rightarrow 2Al_2O_3\,(s)$

9. $Ca\,(s) + 2H_2O\,(l) \rightarrow Ca(OH)_2\,(aq) + H_2\,(g)$

10. $Cr_2O_7^{2-}\,(aq) + 14H^+\,(aq) + 6I^-\,(aq) \rightarrow 2Cr^{3+}\,(aq) + 7H_2O\,(l) + 3I_2\,(aq)$

11. $CuO\,(s) + 2HCl\,(aq) \rightarrow CuCl_2\,(aq) + H_2O\,(l)$

12. $AgNO_3\,(aq) + NaCl\,(aq) \rightarrow AgCl\,(s) + NaNO_3\,(aq)$

13. $SnCl_2\,(aq) + HgCl_2\,(aq) \rightarrow Hg\,(l) + SnCl_4\,(aq)$

14. $2Na_2S_2O_3\,(aq) + I_2\,(aq) \rightarrow 2NaI\,(aq) + Na_2S_4O_6\,(aq)$

15. $2Fe(NO_3)_3\,(aq) + 2KI\,(aq) \rightarrow 2Fe(NO_3)_2\,(aq) + 2KNO_3\,(aq) + I_2\,(aq)$

Chemistry in Society

Test 3.20 Writing ion-electron equations

The questions in this test refer to reactions that occur during redox processes.

Decide
(i) the number of H^+ (aq) ions
(ii) the number of electrons, e^-
required to balance each of the ion-election equations.

A. 1	B. 2	C. 3	D. 4
E. 5	F. 6	G. 7	H. 8
I. 10	J. 12	K. 14	L. 16

For each of the questions, **two** answers should be given.

1. SO_3^{2-} → SO_4^{2-}

2. XeO_3 → Xe

3. ClO^- → Cl_2

4. IO_3^- → I_2

5. PbO_2 → Pb^{2+}

6. $Cr_2O_7^{2-}$ → $2Cr^{3+}$

7. MnO_2 → Mn^{2+}

8. NO_3^- → NO

9. MnO_4^- → Mn^{2+}

10. H_2O_2 → H_2O

11. ClO_4^- → Cl_2

Test 3.21 — Neutralisation (revision)

Questions 1 to 6 refer to the reaction of 20 cm³ of sodium hydroxide solution, concentration 0.2 mol l⁻¹, with acids.

Decide whether the reaction of each of the acids

    A. results in a neutral solution
    B. does **NOT** result in a neutral solution.

1. 20 cm³ hydrochloric acid, concentration 0.2 mol l⁻¹

2. 40 cm³ hydrochloric acid, concentration 0.1 mol l⁻¹

3. 10 cm³ hydrochloric acid, concentration 0.2 mol l⁻¹

4. 20 cm³ sulphuric acid, concentration 0.2 mol l⁻¹

5. 20 cm³ sulphuric acid , concentration 0.1 mol l⁻¹

6. 10 cm³ sulphuric acid, concentration 0.2 mol l⁻¹

Questions 7 to 12 refer to the reaction of 10 cm³ of hydrochloric acid, concentration 1 mol l⁻¹, with excess of sodium carbonate.

Decide whether the reaction of each of the acids

    A. gives the same final volume of carbon dioxide
    B. does **NOT** give the same final volume of carbon dioxide.

7. 20 cm³ of hydrochloric acid, concentration 2 mol l⁻¹

8. 50 cm³ of hydrochloric acid, concentration 0.2 mol l⁻¹

9. 20 cm³ of sulphuric acid, concentration 2 mol l⁻¹

10. 40 cm³ of hydrochloric acid, concentration 0.1 mol l⁻¹

11. 10 cm³ of sulphuric acid, concentration 0.5 mol l⁻¹

12. 20 cm³ of sulphuric acid, concentration 1 mol l⁻¹

13. What volume of hydrochloric acid, concentration 0.1 mol l^{-1}, is required to neutralise 100 cm^3 of sodium hydroxide solution, concentration 0.1 mol l^{-1}?

    A. 25 cm^3 B. 50 cm^3 C. 100 cm^3 D. 200 cm^3

14. What volume of sodium hydroxide solution, concentration 0.5 mol l^{-1}, will be neutralised by 50 cm^3 of sulphuric acid, concentration 0.2 mol l^{-1}?

    A. 40 cm^3 B. 50 cm^3 C. 100 cm^3 D. 200 cm^3

15. If 100 cm^3 of nitric acid is neutralised by 50 cm^3 of potassium hydroxide solution, concentration 0.2 mol l^{-1}, what is the concentration of the acid in mol l^{-1} ?

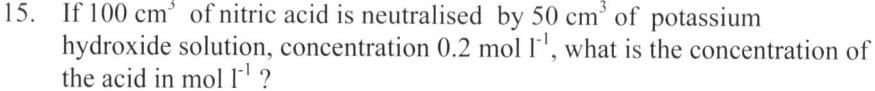

    A. 0.1   B. 0.2   C. 0.4   D. 0.5

16. What volume of sodium hydroxide solution, concentration 1 mol l^{-1}, will be neutralised by 50 cm^3 of hydrochloric acid, concentration 0.5 mol l^{-1}?

    A. 10 cm^3 B. 25 cm^3 C. 50 cm^3 D. 100 cm^3

17. If 20 cm^3 of potassium hydroxide solution is neutralised by 50 cm^3 of sulphuric acid, concentration 0.1 mol l^{-1}, what is the concentration of the alkali?

    A. 0.1   B. 0.2   C. 0.4   D. 0.5

18. What volume of sulphuric acid, concentration 0.1 mol l^{-1}, is required to neutralise 10 cm^3 of lithium hydroxide solution, concentration 0.5 mol l^{-1}?

    A. 5 cm^3 B. 10 cm^3 C. 25 cm^3 D. 50 cm^3

Test 3.22 — Redox titrations

1. The ion-electron equations for a redox reaction are:

   $Fe^{2+}(aq) \rightarrow Fe^{3+}(aq) + e^-$
   $Cr_2O_7^{2-}(aq) + 14H^+(aq) + 6e^- \rightarrow 2Cr^{3+}(aq) + 7H_2O(l)$

   How many moles of iron(II) ions are oxidised by one mole of dichromate ions?

   A. 0.17   B. 0.33   C. 1   D. 6

2. The ion-electron equations for a redox reaction are:

   $2I^-(aq) \rightarrow I_2(aq) + 2e^-$
   $MnO_4^-(aq) + 8H^+(aq) + 5e^- \rightarrow Mn^{2+}(aq) + 4H_2O(l)$

   How many moles of iodide ions are oxidised by one mole of permanganate ions?

   A. 0.4   B. 1   C. 2.5   D. 5

3. Dichromate ions react with ethanol in acidic solution.

   $2Cr_2O_7^{2-}(aq) + 3C_2H_5OH(aq) + 16H^+(aq) \rightarrow 3CH_3COOH(aq) + 4Cr^{3+}(aq) + 11H_2O(l)$

   How many moles of ethanol are oxidised by one mole of dichromate ions?

   A. 1   B. 1.5   C. 2   D. 3

4. Permanganate ions react with hydrogen peroxide in acidic solution.

   $2MnO_4^-(aq) + 6H^+(aq) + 5H_2O_2(l) \rightarrow 2Mn^{2+}(aq) + 8H_2O(l) + 5O_2(g)$

   How many moles of hydrogen peroxide will react completely with 100 cm³ of permanganate solution, concentration 0.1 mol l⁻¹?

   A. 0.01   B. 0.025   C. 0.05   D. 0.1

5. Iron (II) ions react with dichromate ions in acidic solution.

$$6Fe^{2+}(aq) + Cr_2O_7^{2-}(aq) + 14H^+(aq) \rightarrow 6Fe^{3+}(aq) + 2Cr^{3+}(aq) + 7H_2O(l)$$

If 25 cm³ of dichromate solution reacted with 60 cm³ of iron(II) ion solution, concentration 0.1 mol l⁻¹, calculate the concentration of the dichromate solution, in mol l⁻¹.

A. 0.01   B. 0.02   C. 0.04   D. 0.06

6. The concentration of hydrogen peroxide solution can be found by a redox titration with acidified potassium permanganate solution. The equation for the reaction which takes place is:

$$5H_2O_2(aq) + 2MnO_4^-(aq) + 6H^+(aq) \rightarrow 5O_2(g) + 2Mn^{2+}(aq) + 8H_2O(l)$$

It was found that 20.0 cm³ of hydrogen peroxide solution reacted with 40 cm³ of 0.02 mol l⁻¹ potassium permanganate solution when titrated.

Calculate the concentration of the hydrogen peroxide solution, in mol l⁻¹.

A. 0.1   B. 0.8   C. 2   D. 4

7. The chlorine levels in swimming pools can be determined by titrating samples against acidified iron(II) sulphate solution. The reaction taking place is:

$$Cl_2(aq) + 2Fe^{2+}(aq) \rightarrow 2Cl^-(aq) + 2Fe^{3+}(aq)$$

A 100 cm³ sample of water from a swimming pool required 25 cm³ of iron(II) sulphate concentration 2.0 mol l⁻¹, to reach the end point.

Calculate the chlorine concentration, in g l⁻¹, in the swimming pool water.

A. 0.25   B. 17.75   C. 35.5   D. 71

Test 3.23   Chromatography

Questions 1 and 3 relate to gas chromatography.

Decide whether each of the statements is

    A.   TRUE   **B.**   FALSE.

1. Molecules with a higher molecular mass (assuming similar polarity) have a greater retention time.

2. More polar molecules (assuming similar molecular mass) have a greater retention time.

3. A carbonyl group ($C=O$) is more polar than a hydroxyl group (-OH).

4. A mixture of limonene and humulene is injected into a gas/liquid chromatograph.

The chromatogram obtained is shown.

Which terpene corresponds to peak **X**?

    A.   limonene
    B.   humulene

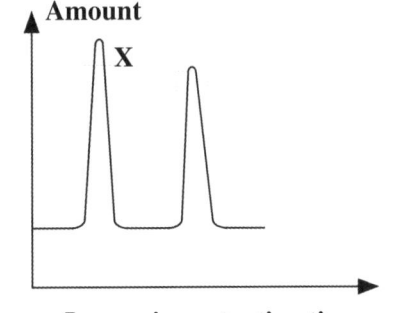

5. A mixture of two alcohols **X** and **Y** is injected into a gas/liquid chromatograph. The mixture consists of 2 volumes of **X** to 1 volume of **Y**.

The chromatogram obtained is shown.

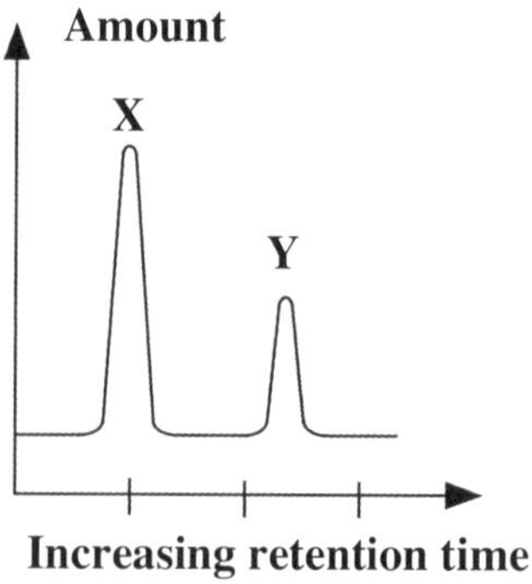

A second chromatogram was obtained using a mixture consisting of the same volume of **X** but a decreased volume of **Y**.

Which chromatogram could have been obtained?

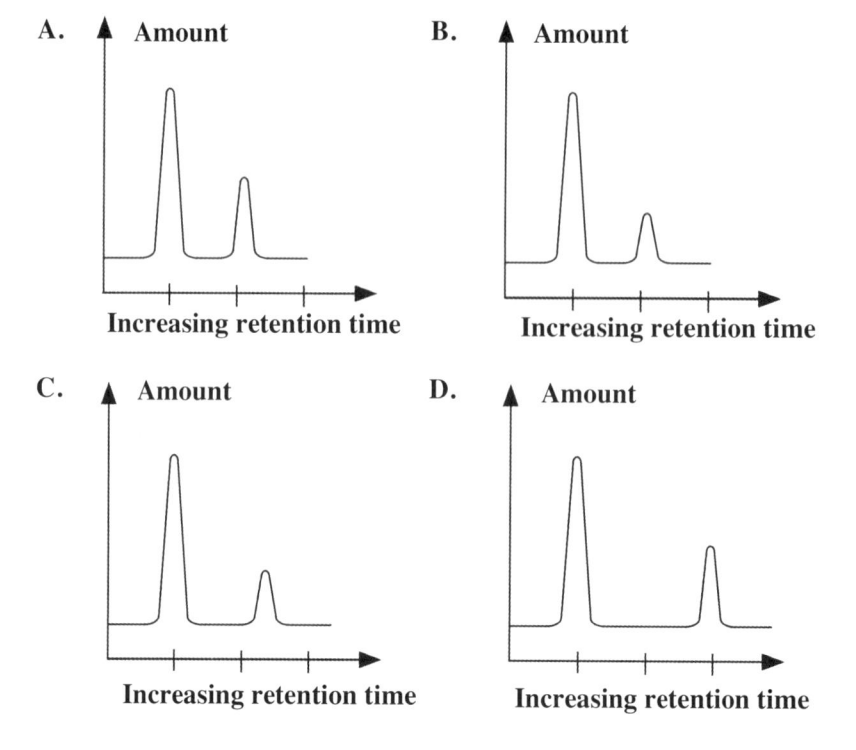

6. A mixture of ammonia and methane is injected into a gas/liquid chromatograph.

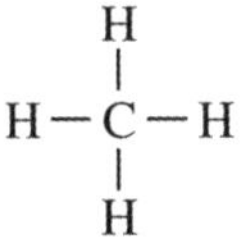

methane

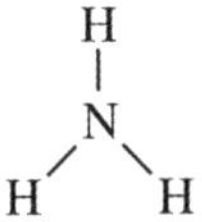

ammonia

The chromatogram obtained is shown.

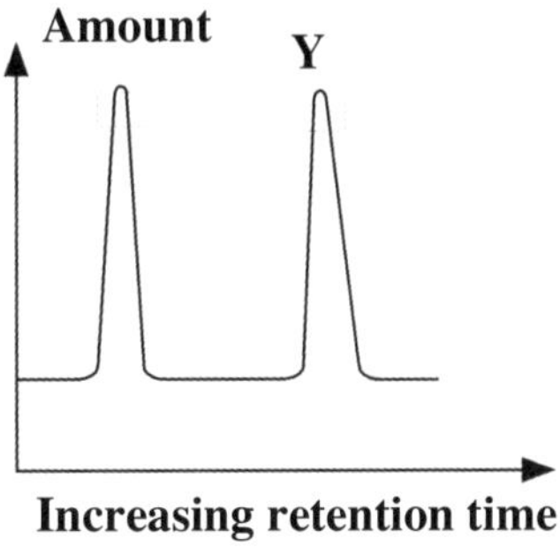

Amount
Y
Increasing retention time

Which gas corresponds to peak **Y**?

A. methane
B. ammonia

Questions 7 to 9 refer to the paper chromatogram used to identify amino acids in unknown samples **P, Q, R** and **S**. The names of the known amino acids are in the table.

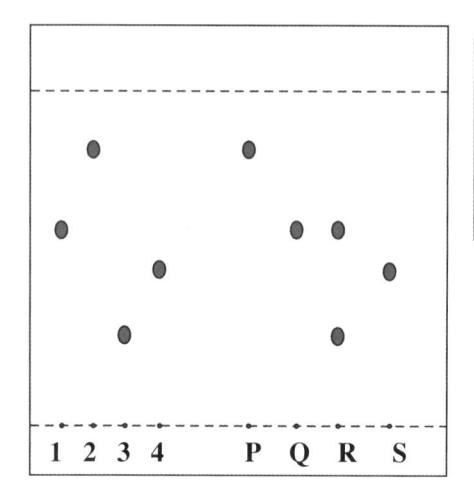

No	Amino acid
1	asparagine
2	phenylalanine
3	glycine
4	tyrosine

7. Which of the four known amino acids is likely to have the **least** polar structure?

    A. asparagine
    B. phenylalanine
    C. glycine
    D. tyrosine

8. Which of the four known amino acids is found in sample **Q**?

    A. asparagine
    B. phenylalanine
    C. glycine
    D. tyrosine

9. Which of the four unknown samples is **not** pure?

    A. P
    B. Q
    C. R
    D. S

Test 3.24   Practical skills

In questions 1 to 11 decide whether the statement is

    A.   TRUE   B.   FALSE.

1. A 25 cm^3 pipette can be used to measure out 20 cm^3 of solution.

2. A 50 cm^3 burette can be used to measure out 20 cm^3 of solution.

3. A 100 cm^3 measuring cylinder can be used to measure out 50 cm^3 of solution.

4. The most accurate measurement (lower error) for the volume of a liquid is obtained using a pipette.

5. The correct method for filling a pipette is to draw in the solution, slowly at the end, until the bottom of the meniscus touches the mark.

6. The correct volume for a filled pipette is obtained when the top of the meniscus is in line with the mark.

7. A standard solution always has a concentration of 1 mol l^{-1}.

8. A standard solution is made by adding a known mass of solid to a conical flask and making up to the mark with water.

9. The solid used to make a standard solution must be dry.

10. The mass of solid used to make a standard solution is always 10 g.

11. The solid used to make a standard solution must be pure.

12. The final step in making a standard solution involves using a dropping pipette to remove drops of solution until the bottom of the meniscus is on the mark.

13. It is more accurate (lower error in measurement) to make a standard solution with a very low concentration, e.g. 1 mg l⁻¹, by measuring out the required mass of solid and making up to the mark than by diluting a solution with a higher concentration.

14. A 0.10 mol l⁻¹ solution could be prepared most accurately from a 1.0 mol l⁻¹ solution using

    A. a 1 cm³ dropping pipette and a 10 cm³ measuring cylinder
    B. a 10 cm³ measuring cylinder and a 100 cm³ volumetric flask
    C. a 25 cm³ pipette and a 250 cm³ volumetric flask
    D. a 50 cm³ burette and a 500 cm³ measuring cylinder.

In questions 15 to 18 decide whether each of the salts

A. can be separated from a reaction mixture by filtration
B. CANNOT be separated from a reaction mixture by filtration.

You may wish to use the Data Booklet.

15. sodium nitrate

16. silver chloride

17. potassium sulphate

18. magnesium carbonate

19. A student carried out an investigation to measure the acidity of a water sample. The results of the three titrations are shown.

First titration/cm³	Second titration/cm³	Third titration/cm³
25.7	24.5	24.3

The volume used from the results in the calculation would be

A. 25.7 **B.** 24.5 **C.** 24.3 **D.** 24.4.

20. Four technicians carried out the experiment to measure the percentage of oxygen in a sample of air.

	Experiment 1	Experiment 2	Experiment 3
Technician 1	18.7 %	19.2 %	18.3 %
Technician 2	19.4 %	19.6 %	19.5 %
Technician 3	19.7 %	19.5 %	19.3 %
Technician 4	18.1 %	19.1 %	20.1 %

The most reproducible results were obtained by

A. Technician 1
B. Technician 2
C. Technician 3
D. Technician 4.

21. Sulphur dioxide is denser than air.

Identify the most suitable arrangement for collecting a sample of sulphur dioxide.

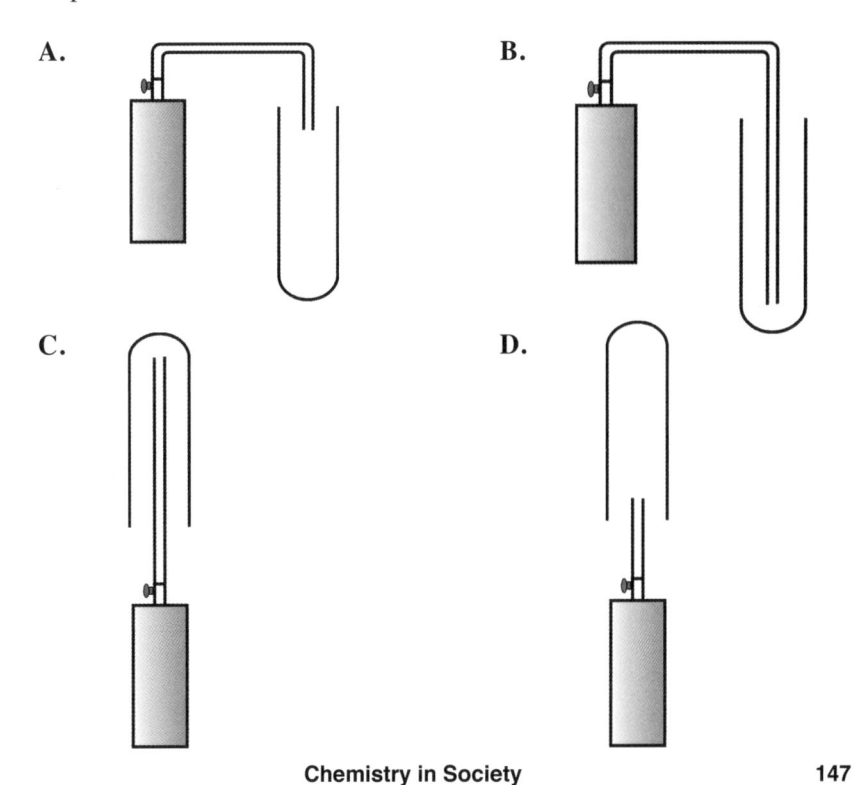

A.

B.

C.

D.

Answers   Unit 1 Chemical Changes and Structure

Test 1.1

1. A
2. B
3. A
4. A
5. B
6. A
7. A
8. A
9. B
10. A
11. B
12. A
13. B
14. B
15. C
16. A
17. C
18. C

Test 1.2

1. A
2. B
3. A
4. B
5. A
6. A
7. A
8. B
9. B
10. A
11. A
12. C
13. D
14. B
15. B
16. B
17. A
18. B

Test 1.3

1. C
2. D
3. A
4. B
5. B
6. C
7. D
8. A
9. B
10. C
11. A
12. A
13. A
14. C
15. A
16. B
17. C
18. A
19. D
20. C
21. B
22. D

Test 1.4

1. A
2. E
3. A
4. C
5. A
6. D
7. C
8. A
9. E
10. C
11. A
12. E
13. D
14. C
15. A
16. B
17. A
18. C
19. D

Test 1.5

1. A
2. B
3. B
4. B
5. A
6. A
7. B
8. B
9. B
10. D

Test 1.6

1. B
2. A
3. B
4. B
5. A
6. B
7. A
8. A
9. B
10. A
11. B
12. B
13. A
14. A
15. B
16. A
17. A
18. A
19. A
20. B
21. B
22. B
23. B
24. B
25. B
26. A
27. A
28. A
29. A
30. B
31. A
32. B
33. A

Test 1.7

1. A
2. B
3. A
4. B
5. A
6. B
7. B
8. A
9. A
10. A
11. C
12. A
13. C
14. A
15. A
16. C
17. C
18. C
19. B

Test 1.8

1. B
2. B
3. A
4. B
5. B
6. C
7. B
8. C
9. D
10. D
11. B
12. C
13. D
14. D

Test 1.9

1. A
2. C
3. B
4. A
5. C
6. B
7. A
8. C
9. A
10. A
11. B
12. C
13. E
14. B
15. B
16. C
17. A
18. A

Test 1.10

1. B
2. A
3. B
4. A
5. A
6. B
7. A
8. A
9. A
10. B
11. A
12. C
13. B
14. B
15. B
16. C
17. B
18. A
19. C

Test 1.11

1. C
2. B
3. A
4. B
5. D
6. B
7. C
8. C
9. B
10. A
11. D
12. A
13. B
14. C
15. D
16. B
17. E
18. C
19. D
20. E
21. A
22. E
23. E
24. A
25. D

Test 1.12

1. B
2. D
3. D
4. C
5. B
6. A
7. D
8. A
9. B
10. C
11. D
12. A
13. A

Test 1.13

1. B
2. A
3. B
4. B
5. C
6. B
7. C
8. A
9. C
10. B
11. C
12. A
13. D
14. A
15. D
16. A
17. D
18. B

Test 1.14

1. BDG
2. ADE
3. ACG
4. ACF
5. ADG
6. ACE
7. BDE
8. ACE
9. ACE
10. BDG
11. C
12. D

Test 1.15

1. B
2. C
3. C
4. A
5. A
6. D
7. C
8. A
9. D
10. A
11. C
12. C
13. A
14. B
15. D
16. A
17. C
18. A
19. B
20. D
21. B
22. D
23. D

Answers

Answers

Unit 2 Nature's chemistry

Test 2.1

1. AC
2. AC
3. BD
4. BD
5. AC
6. AC
7. BD
8. BD
9. AC
10. BD
11. A
12. I
13. D
14. H
15. J
16. C
17. B
18. B
19. A
20. C
21. C
22. C
23. A
24. D
25. D
26. D
27. C
28. A
29. C
30. A
31. B

Test 2.2

1. A
2. G
3. K
4. H
5. E
6. L
7. F
8. J
9. O
10. R
11. D
12. A
13. F
14. B
15. I
16. C
17. G
18. L
19. A
20. C

Test 2.3

1. B
2. A
3. B
4. B
5. B
6. A
7. B
8. A
9. A
10. B
11. B
12. B
13. A
14. B
15. B
16. A
17. A
18. B
19. A
20. A
21. A
22. B

Test 2.4

1. CMQS
2. FP
3. BIJNV
4. AGOR
5. DHKTU

Test 2.5

1. A
2. B
3. A
4. B
5. D
6. B
7. C
8. A

Test 2.6

1. B
2. A
3. A
4. A
5. B
6. A
7. A
8. C
9. D
10. A
11. B
12. A
13. B
14. B
15. A
16. A
17. B

Test 2.7

1. B
2. A
3. B
4. A
5. A
6. B
7. B
8. A
9. B
10. B
11. B
12. FH
13. DFH
14. C
15. B
16. C
17. D
18. B

Test 2.8

1. BF
2. B
3. B
4. A
5. C
6. C
7. C
8. B

Test 2.9

1. EFGLN
2. ACHIKMO
3. BDJP

Test 2.10

1. A
2. B
3. A
4. C
5. A
6. C
7. A
8. B
9. A
10. B
11. C
12. B
13. A
14. B
15. A
16. B
17. A
18. B
19. A
20. B
21. A
22. B

Test 2.11

1. B
2. B
3. A
4. B
5. A
6. B
7. C
8. B
9. D
10. A
11. B
12. D
13. A
14. A
15. B
16. B
17. A
18. A
19. A
20. B
21. A
22. A
23. B
24. B

Test 2.12

1. B
2. A
3. B
4. A
5. A
6. B
7. B
8. C
9. D
10. D

Test 2.13

1. AE
2. D
3. I
4. D
5. AB
6. C
7. H
8. D
9. G
10. F
11. D
12. A
13. D
14. I
15. G
16. D
17. C
18. H
19. AB
20. F

Test 2.14

1. B
2. B
3. C
4. B
5. C
6. A
7. D
8. D
9. B
10. A

Test 2.15

1. A
2. B
3. A
4. A
5. B
6. B
7. C
8. D
9. A
10. B
11. D

Test 2.16

1. A
2. B
3. B
4. A
5. A
6. C
7. A
8. A
9. C
10. C
11. B
12. A
13. C
14. A
15. C
16. D
17. B
18. D

Test 2.17

1. A
2. B
3. A
4. A
5. D
6. D
7. A
8. BDE
9. ACF
10. A
11. B
12. B
13. A
14. C
15. A
16. B
17. A
18. B
19. B
20. A
21. B

Test 2.18	Test 2.19	Test 2.20	Test 2.21	Test 2.22
1. B	1. FG	1. A	1. B	1. A
2. A	2. DG	2. A	2. B	2. B
3. D	3. AC	3. B	3. B	3. A
4. D	4. AE	4. B	4. A	4. A
5. A	5. BD	5. A	5. B	5. B
6. A	6. DG	6. B	6. A	6. B
7. D	7. AB	7. A	7. C	7. A
8. A	8. FG	8. B	8. A	8. B
9. B		9. B	9. B	9. B
10. C		10. A	10. C	10. A
11. B		11. B	11. B	11. B
12. B		12. D	12. A	12. C
13. A		13. C	13. E	13. B
14. D			14. D	14. C
			15. C	15. A
				16. B
				17. C
				18. A
				19. BE
				20. CDF

Answers

Unit 3 Chemistry in Society

Test 3.1	Test 3.2		Test 3.3		Test 3.4
1. B	1. A	12. D	1. B	14. A	1. D
2. B	2. A	13. C	2. A	15. A	2. B
3. D	3. A	14. D	3. A	16. A	3. C
4. D	4. B	15. D	4. A	17. A	4. A
5. D	5. B	16. D	5. B	18. A	5. A
6. A	6. C	17. B	6. B	19. B	6. B
7. C	7. B	18. A	7. B	20. A	7. D
8. C	8. D	19. D	8. A	21. A	8. C
	9. B	20. A	9. B	22. B	9. C
	10. B	21. A	10. B	23. A	10. C
	11. C	22. D	11. B	24. B	11. A
			12. A	25. A	12. A
			13. A	26. B	13. A
					14. B

Test 3.5	Test 3.6	Test 3.7	Test 3.8	
1. A	1. B	1. A	1. C	13. C
2. C	2. A	2. B	2. B	14. A
3. D	3. B	3. C	3. A	15. B
4. C	4. B	4. C	4. B	16. B
5. B	5. C	5. A	5. A	17. C
6. D	6. A	6. A	6. B	18. A
7. B	7. C	7. A	7. B	19. B
8. C	8. C	8. C	8. C	20. C
9. B	9. A	9. C	9. B	21. C
10. C	10. A	10. A	10. B	22. B
	11. B	11. B	11. A	23. A
	12. D	12. A	12. A	24. A
	13. A	13. B		
	14. D	14. B		
	15. C	15. B		
	16. C	16. B		
	17. A	17. A		
	18. A			

Test 3.9	Test 3.10	Test 3.11	Test 3.12	Test 3.14
1. B	1. A	1. A	1. A	1. B
2. A	2. B	2. C	2. A	2. A
3. B	3. A	3. C	3. C	3. B
4. A	4. B	4. A	4. D	4. B
5. B	5. B	5. C	5. A	5. C
6. A	6. C	6. B	6. C	6. A
7. B	7. A	7. A		7. A
8. A	8. A	8. C		8. A
9. A	9. C	9. C		9. C
10. B	10. B	10. C	Test 3.13	10. D
11. B	11. C	11. B		
12. B	12. A	12. C	1. D	
13. A	13. C	13. D	2. D	
14. B	14. B	14. C	3. A	
15. A	15. A		4. C	
16. B	16. C		5. C	
17. B	17. B		6. C	
18. D	18. C		7. C	
	19. A		8. A	
	20. B		9. B	
	21. B		10. D	
	22. A		11. D	

Test 3.15	Test 3.16	Test 3.17	Test 3.18	Test 3.19
1. D	1. AC	1. B	1. B	1. A
2. A	2. AD	2. A	2. A	2. B
3. A	3. BC	3. A	3. A	3. B
4. A	4. BC	4. B	4. A	4. A
5. C	5. BD	5. A	5. B	5. A
6. C	6. AD	6. A	6. A	6. B
7. A	7. A	7. A	7. B	7. B
8. C	8. D	8. B	8. B	8. A
9. C	9. B	9. A	9. A	9. A
10. B	10. B	10. B	10. A	10. A
11. B	11. A	11. A		11. B
12. D	12. B	12. B		12. B
13. A	13. B			13. A
	14. B			14. A
	15. A			15. A
	16. C			
	17. A			

Test 3.20	Test 3.21	Test 3.22	Test 3.23	Test 3.24
1. BB	1. A	1. D	1. A	1. B
2. FF	2. A	2. D	2. A	2. A
3. DB (or BA)	3. B	3. B	3. B	3. A
4. JI (or FE)	4. B	4. B	4. A	4. A
5. DB	5. A	5. C	5. C	5. B
6. KF	6. A	6. A	6. B	6. B
7. DB	7. B	7. B	7. B	7. B
8. DC	8. A		8. A	8. B
9. HE	9. B		9. C	9. A
10. BB (or AA)	10. B			10. B
11. LJ (or HF)	11. A			11. A
	12. A			12. B
	13. C			13. B
	14. A			14. C
	15. A			15. B
	16. B			16. A
	17. D			17. B
	18. C			18. A
				19. D
				20. B
				21. B